MARINES

MSgt Andy Bufalo USMC (Ret)

All Rights Reserved. Copyright © 2010 by
All American Books, an imprint of S&B Publishing.

ISBN 978-0-9817007-9-3

This book may not be reproduced in whole or in part without the written permission of the author or the publisher.

First Printing – March 2010
Printed in the United States of America

www.AllAmericanBooks.com

Reel Marines

Reel Marines

OTHER BOOKS BY ANDY BUFALO

HOLLYWOOD MARINES
Celebrities Who Served in the Corps

SWIFT, SILENT & SURROUNDED
Sea Stories and Politically Incorrect Common Sense

THE OLDER WE GET, THE BETTER WE WERE
MORE Sea Stories and Politically Incorrect Common Sense
Book II

NOT AS LEAN, NOT AS MEAN, STILL A MARINE!
Even MORE Sea Stories and Politically Incorrect Common Sense
Book III

EVERY DAY IS A HOLIDAY…
Every Meal is a Feast!
Yet Another Book of Sea Stories and Politically Incorrect Common Sense
Book IV

THE ONLY EASY DAY WAS YESTERDAY
Marines Fighting the War on Terrorism

HARD CORPS
The Legends of the Marine Corps

AMBASSADORS IN BLUE
In Every Clime and Place
Marine Security Guards Protecting Our Embassies Around the World

THE LORE OF THE CORPS
Quotations By, For & About Marines

Reel Marines

"The Marine Corps is the Navy's police force, and as long as I am President that is what it will remain. They have a propaganda machine that is almost equal to Stalin's." – President Harry S. Truman

Reel Marines

This book is dedicated to every Marine who has ever appeared on the silver screen, both real and fictional, and to the legendary exploits which have made the United States Marine Corps the subject of so many tales.

TABLE OF CONTENTS

A Few Good Men 1
A Rumor of War 8
All the Young Men 12
Ambush Bay 17
An Officer and a Gentleman 21
Battle Cry 29
Beachhead 34
Born on the Fourth of July 39
Death Before Dishonor 45
Ears, Open. Eyeballs, Click. 48
First to Fight 51
Flags of Our Fathers 54
Flying Leathernecks 61
Full Metal Jacket 65
Generation Kill 75
Guadalcanal Diary 79
Gung Ho! 83
Halls of Montezuma 88
Heartbreak Ridge 93
Independence Day 100
Jarhead 108
Major Payne 114
Pride of the Marines 118
Purple Hearts 121
Retreat Hell 124
Rules of Engagement 130

Reel Marines

Salute to the Marines	136
Sands of Iwo Jima	141
Shooter	146
Sniper	153
Space	157
Taking Chance	161
Tell It to the Marines	166
The Boys in Company C	170
The D.I.	173
The Great Santini	176
The Marine	180
The Outsider	185
The Pacific	188
The Rock	193
The Shores of Tripoli	199
The Siege of Firebase Gloria	204
To the Shores of Hell	207
To the Shores of Iwo Jima	210
Tribes	213
Uncommon Valor	216
Wake Island	219
What Price Glory?	223
Wind and the Lion	226
Windtalkers	232
55 Days at Peking	236
101 More Movies	241
Photo Gallery	265

Reel Marines

Beginning with *The Star Spangled Banner* way back in 1918, the Marine Corps discovered the use of motion pictures. In exchange for a favorable portrayal that stimulated recruiting and gave an impressive view to the public and Congress, the Marines provided uniformed extras, locations, equipment, and technical advisers that provided their expertise to the producers. In 1926 MGM's *Tell it to the Marines* and Fox's *What Price Glory?* nearly led to a court battle to see whether one studio could copyright the Marines to prevent other films from being made. Over the years Camp Pendleton has been dressed up to represent Central American nations, China, Pacific islands, and New Zealand. Joseph H. Lewis's *Retreat Hell* (1952) and Allan Dwan's John Payne film *Hold Back The Night* (1956) had the California base covered in studio snow with their hills and roads painted white. Camp Pendleton later doubled as Vietnam in Marshall Thompson's *To the Shores of Hell* (1966), and while filming *Battle Cry* at the base in 1954 Raoul Walsh's Marine technical adviser said he had joined the Corps after seeing Walsh's *What Price Glory?* Marine boot camp was also depicted in Jack Webb's *The D.I.* (1957) and Stanley Kubrick's *Full Metal Jacket* and led to more enlistments.

Perhaps the biggest and most important Marine film was Allan Dwan's John Wayne epic *Sands of Iwo Jima* in 1949. At the time the Corps was facing its toughest battle for survival with Congressional hearings being conducted on

Reel Marines

significantly reducing the size and capabilities of the Marines. The Corps spared no expense in providing Marines, equipment, and combat cameraman footage - and even trained the actors to look, dress, and act like Marines. The movie trailer told the viewer "these are YOUR boys" and the large scale battle scenes impressed the public. The ending - with Wayne's death, John Agar growing into manhood and taking Wayne's place, the depiction of the flag raising on Iwo Jima, and Agar leading his squad into the mists of history as a choral version of *The Marine's Hymn* begins playing quietly and grows louder in the background - led to an unforgettable image of the Corps. Congressmen voting on appropriations for would forever be deluged with letters from their constituents who had images of Marines running out of ammunition and being treacherously bayoneted by Japanese soldiers. President Harry S. Truman, who had himself served as an Army artillery officer during World War I, once said the Corps had "a propaganda machine bigger than Joseph Stalin's" but had to apologize when he presented the 1st Marine Division with a Presidential Unit Citation for their gallantry during the Korean War.

The Marine Corps, as well as individual Marines, have been depicted many ways over the years and from many different points of view. Gunny Highway, Bull Meecham and Colonel Nathan Jessup are perceived as caricatures by many unitiated viewers, but are at the same time they are embraced by real Marines as icons of our warrior culture. When Highway tells his Commanding Officer "it's a clusterfuck, sir" we all nod our heads in understanding. When Bull Meechaam laments the lack of a war for him to fight, we all feel his pain. And when Colonel Jessup tells the hotshot Navy lawyer played by Tom Cruise - who is wearing

Reel Marines

his "faggoty white uniform" - that we "want him on that wall... need him on that wall," we know he speaks the truth.

There are, of course, many memorable celluloid Marines who are not included in this book because I felt compelled to limit myself to movies which are "about" the Corps. Al Pacino's turn as war hero Captain Michael Corleone at the beginning of the *Godfather* immediately comes to mind, as does Wesley Snipes' portrayal of framed Marine/CIA operative Mark Warren in *U.S. Marshals*.

By the time you read this there will no doubt be more movies to add to the list, but that will in no way detract from the Corps' history according to Hollywood up to this point. Any way you slice it a lot of films have been made about the Corps, with the fifty-one profiled in this book being, in my opinion, the most important - and entertaining.

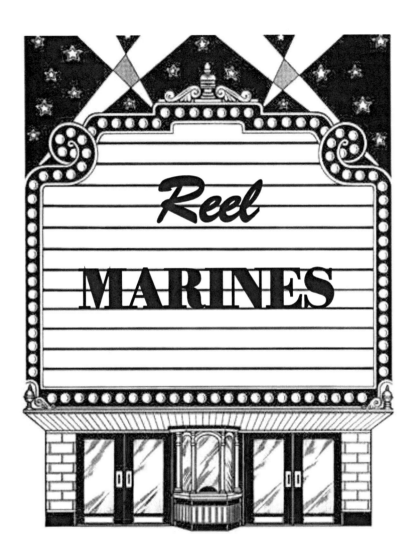

Reel Marines

A FEW GOOD MEN

Release date: December 11, 1992
Running time: 138 minutes
Historical context: Peacetime

Tagline: In the heart of the nation's capital, in a courthouse of the U.S. government, one man will stop at nothing to keep his honor, and one will stop at nothing to find the truth.

Cast

Tom Cruise - Lieutenant (j.g.) Daniel Kaffee
Demi Moore - Lieutenant Commander JoAnne Galloway
Jack Nicholson - Colonel Nathan Jessup
Kevin Pollak - Lieutenant (USN) Sam Weinberg
Kevin Bacon - Captain Jack Ross
J. T. Walsh - Lieutenant Colonel Markinson
Kiefer Sutherland - First Lieutenant Kendrick

Quote: "You can't handle the truth!" – Colonel Nathan Jessup

Highlights

This film is a modern morality play, and it's obvious the writers didn't truly understand the Marine Corps. Even so, Jack Nicholson's iconic performance as Colonel Jessup is one with which every Marine can identify - he is an officer who takes his duty seriously, and he wants to prepare his Marines for combat. Just hearing him tell Tom Cruise "You can't candle the truth" is reason enough to see it!

Reel Marines

Plot

Lieutenant Daniel Kaffee, the son of a former Attorney General and U.S. Navy Judge Advocate General, is an inexperienced Navy JAG lawyer who leads the defense in the court-martial of two Marines when Private Louden Downey and Lance Corporal Harold Dawson are accused of murdering a Marine in their unit, Private William Santiago, at the Guantanamo Bay Naval Base in Cuba.

Santiago compared unfavorably to his fellow Marines, had poor relations with them, failed to respect the chain of command, and even went above his superiors to bargain for a transfer in exchange for blowing the whistle on Dawson for firing a possibly illegal shot towards the Cuban side of the island.

In a flashback Colonel Nathan Jessup, the Commanding Officer of the accused, reads the letter detailing the incident to two of his officers - his executive officer, Lieutenant Colonel Markinson, and Santiago's platoon commander, Lieutenant Kendrick. Jessup and Kendrick are incensed at Santiago's actions, but decide not to transfer him despite the objection of Markinson. After giving Markinson a dressing-down for questioning his views on the matter, Jessup calls Kendrick in to discuss "young William's training."

When Dawson and Downey are later arrested for Santiago's murder Naval investigator and lawyer Lieutenant Commander JoAnne Galloway suspects they were carrying out a "Code Red," which is a euphemism for a violent extrajudicial punishment.

Galloway requests to defend them but the case is given to Kaffee, who has a reputation for arranging plea bargains. Colonel Jessup is due to take up an important post at the National Security Agency and it is implied people in high

Reel Marines

places want the case settled with the minimum amount of fuss because of this, however Galloway successfully argues her point of view to Kaffee after Dawson and Downey state they were ordered by Kendrick to shave Santiago's head minutes after Kendrick ordered the platoon not to touch the would-be victim. His death was actually caused when a rag was shoved into his mouth as a gag.

Goaded by Galloway and Dawson, Kaffee agrees to be lead counsel for the defense. Despite initial friction between the two lawyers - since she believes he negotiates plea bargains to avoid having to argue in court, and he believes she is interfering with his handling of the case - their relationship strengthens as the trial progresses, as does Kaffee's effectiveness as a lawyer.

Nonetheless Kaffee believes the case will be lost, and after weeks of intensive preparation he concludes that "We're gonna get creamed!" The prosecution, led by his friend Captain Jack Ross, has a good case since Dawson and Downey do not deny assaulting Santiago and only reveal details crucial to their defense under intense prompting. Downey is a simple-minded young man who is seemingly unaware of the gravity of his situation, while the more authoritative Dawson is determined to complete the trial rather than dishonor himself and the Marine Corps with a plea bargain. When Kaffee negotiates a deal which could reduce their sentences from twenty years to just six months, Dawson rejects it as dishonorable and calls Kaffee a coward for suggesting such a deal - and intentionally fails to salute him when he leaves the room.

In the course of the trial it is established that Code Reds are standard practice in Guantanamo Bay as a means of enforcing discipline and getting sloppy Marines to shape up. Kaffee especially goes after Kendrick, partly because the lieutenant had once denied Dawson a promotion when the latter helped out a fellow Marine who was being subjected to a Code Red.

Lieutenant Colonel Markinson, who has been missing since the incident, resurfaces in Kaffee's car during the trial. During a previous meeting with Jessup Kaffee was told Santiago was due to be transferred off the base for his own safety, but Markinson reveals there was never any intention of transferring Santiago and that transfer orders were created as part of a cover-up long after his death. Kaffee is unable to find evidence corroborating these claims and announces his intention to have Markinson testify, but rather than publicly dishonor himself and the Marine Corps Markinson sends a letter to Santiago's parents in which he blames his own weakness for the loss of their son. He then dresses in full uniform and commits suicide.

At the same time evidence is found which questions whether Kendrick ordered Dawson and Downey to carry out the Code Red, something the defense had always taken for granted. Galloway is convinced Jessup also ordered the Code Red, and tries to persuade Kaffee to cross-examine him on this point. Kaffee recoils, since there is no proof Jessup was involved and such unfounded accusations could result in his being court-martialed.

After Galloway storms out Kaffee reflects on his late father with co-counsel Sam Weinberg. Weinberg admits that, with the evidence they have, Kaffee's father would never call Jessup to the stand, but also says he would rather have the younger Kaffee as the lawyer for Dawson and Downey. Weinberg pushes his friend to consider whether it is he or Lionel Kaffee who is handling the case, and Daniel Kaffee finally decides to put Jessup on the stand.

Reel Marines

In court Kaffee questions Jessup and produces circumstantial evidence which suggests there was never any intention of transferring Santiago. When this proves insufficient, Kaffee confronts Jessup regarding the incompatibility of his ordering Santiago's transfer - ostensibly for his own safety - with his assertion that he had ordered Santiago was not to be touched, and that his orders were always followed. When Kaffee asks Jessup point-blank, "Did you order the Code Red?" the judge announces he is in contempt of court, but Jessup cannot resist the challenge. When Kaffee exclaims, "I want the truth!" Jessup emphatically declares, "You can't handle the truth!"

Because he defends his country Colonel Jessup does not see why Kaffee, who has never been on the front lines, should even question his methods from "under the blanket of the very freedom I provide," but in the face of a point-blank question from Kaffee he furiously declares that he did in fact order the Code Red. At the prompting of Kaffee and the Judge, prosecutor Ross then places Jessup under arrest.

Kaffee later admits to Ross that he used a little bluff which unnerved Jessup just enough to cause his downfall, and Ross announces that Kendrick will also be arrested. Dawson and Downey are found not guilty of murder and conspiracy to commit murder, but are found guilty of "conduct unbecoming a United States Marine" and are dishonorably discharged. Downey is confused by the sentence, because he is sure that Jessup's confession absolves them from blame, but Dawson points out that they failed to fight for those unable to fight for themselves, like Santiago. As they leave, Kaffee tells Dawson that he doesn't need to wear a patch on his arm to have honor, and Dawson responds by exchanging salutes with the officer.

Reel Marines

Trivia

❖ The term "Code Red" is not used in the Marine Corps. The actual term is "blanket party," which is what Private Pyle receives in *Full Metal Jacket*.

❖ The original play was inspired by an actual incident at Guantanamo Bay. Lance Corporal David Cox and nine other enlisted men tied up a fellow Marine and severely beat him for snitching to the Naval Criminal Investigative Service. Cox was acquitted and later Honorably Discharged, but in 1994 he mysteriously vanished and his bullet-riddled body was found three months later. His murder remains unsolved.

❖ While filming the scene in which Kendrick (played by Kiefer Sutherland) is driving Kaffee's group around the base in a Humvee through two columns of marching Marines, Sutherland had trouble driving the extra wide vehicle and actually hit Marines on multiple takes.

❖ In the scene where Kaffee is watching a baseball game at his home after returning from Cuba, the sports announcer heard on the television is legendary San Diego Padres broadcaster Jerry Coleman, himself a former Marine officer and aviator who served in WWII and Korea.

❖ "Conduct unbecoming a United States Marine" is a fictional violation based on the actual violation of "conduct unbecoming an officer."

❖ The film starts with a recital of *Semper Fidelis* by a Marine Corps marching band and a Silent Drill routine performed by the Texas A&M University Fish Drill Team.

Reel Marines

- Tom Cruise's character of Lieutenant Daniel Kaffee was based on David Iglesias, who was at the time a United States Navy Reserve commander.

- Wolfgang Bodison was working for Rob Reiner on *A Few Good Men* as a location scout when Reiner decided he was perfect for the part of Lance Corporal Harold Dawson. Reiner said he looked like a Marine.

- Nicholson's line "You can't handle the truth!" was voted the twenty-ninth greatest American movie quote of all time by the American Film Institute.

- Nicholson's intense performance of the famous courtroom speech was all shot in one take, meaning it was the first and only time he delivered it on-camera.

- *Shaurya* is a Hindi movie heavily inspired by *A Few Good Men*.

Reel Marines

A RUMOR OF WAR

Release date: 24 September 1980
Running time: Runtime: 195 minutes
Historical context: Vietnam

Tagline: Every generation is doomed to fight its war.

Cast

Brad Davis - Lieutenant Philip Caputo
Keith Carradine - Lieutenant Murph McCoy
Michael O'Keefe - Lieutenant Walter Cohen
Brian Dennehy – Staff Sergeant Ned Coleman
Stacy Keach - Major Ball

Quote: "It was the same feeling we had experienced on the first operation, a sense of being marooned on a hostile shore from which there was no certainty of return." – Lieutenant Caputo

Highlights

A Rumor of War is based upon the book of the same name by real life Vietnam-era Marine Philip Caputo, and its authenticity is derived from the fact that the conduct of warfare is in no way glamorized – and in fact, some of the nonsensical occurrences will cause many old salts to nod their heads knowingly. Watch for Brian Dennehy, who served in the Corps from 1959 to 1963, as Sergeant Ned "Frosen Chosin" Coleman.

Reel Marines

Plot

The book upon which the movie is based is divided into three parts. The first section, *The Splendid Little War,* describes Lieutenant Philip Caputo's personal reasons for joining the Marine Corps, the training that followed, and his eventual arrival in Vietnam. Caputo was a member of the 9th Expeditionary Brigade, the first American regular troops sent to take part in the Vietnam War. He arrived on March 8, 1965 and his early experiences reminded him of the colonial wars portrayed by Rudyard Kipling. The Brigade was initially deployed to Da Nang in a "purely defensive" posture, primarily to set a perimeter around the airstrip which handled the arrival and departure of military goods and personnel. The first skirmishes against the North Vietnamese Army and the Viet Cong left Caputo and his comrades with the impression the Vietnam conflict was small and relatively unimportant.

In the second part of the book, *The Officer in Charge of the Dead,* Caputo is reassigned from his rifle company to a desk job and tasked with documenting casualties. The position in the Joint Staff of the brigade was a change he condemned because he was proud of his rifle company duties, but this distance from the Main Line of Resistance gave Caputo a different perspective of the conflict. Caputo described senior officers as being more worried about trivial matters than strategy, and witnesses enemy corpses being treasured as hunting trophies and shown off to generals. He also sees American corpses which carry evidence of Viet Cong torture.

In the third part, *In Death's Grey Land,* Caputo is reassigned to a rifle company. He describes the North Vietnamese Army and the Viet Cong as fierce and clever

Reel Marines

warriors who have earned the grudging respect of the Americans. Caputo and his fellow Marines have by now stopped wishing for epic, World War II-style battles, and he describes how they learned to detect boobytraps, counter-snipe, and comb the jungle in search of enemy bunkers and rations. Caputo took part in these operations until troops under his command miscarried orders and shot two suspects deliberately, and he assumed full responsibility for the incident and faced a court-martial. Eventually Caputo was relieved of his command, the charges were dropped, and he was reassigned to North Carolina where he received an honorable discharge.

Trivia

- *A Rumor of War* was originally a 1980 television miniseries, based on the 1977 autobiography by Philip Caputo about his service in the Marine Corps in the early years of American involvement in the Vietnam War.

- It was one of the earliest serious works of television or film drama to be based on U.S. combat experience in Southeast Asia.

- Almost ten years after the end of his tour of duty Philip Caputo returned to Vietnam as a war journalist for a newspaper, and old memories of his war experiences and his comrades flooded his mind as he witnessed the fall of Saigon to the troops of North Vietnam. Caputo left Vietnam once again on April 29, 1975.

- A postscript published in 1996 details some of the anxieties Caputo experienced while writing the memoir, and the difficulties he had handling his fame and notoriety after its publication.

Reel Marines

- The miniseries was filmed at Camp Pendleton and Churubusco Studios in Mexico.

- The producers could not find enough Sikorsky HUS-1 helicopters, so they used UH-1 Huey helicopters instead.

- Years later Brian Dennehy gave interviews to *Playboy* magazine and the *New York Times* in which he stated he had served in Vietnam and been wounded, when in reality his four years in the Marine Corps had been during peacetime. Some people have speculated that those misrepresentations grew out of his portrayal of Sergeant Coleman in this film.

Reel Marines

ALL THE YOUNG MEN

Release date: 26 August 1960
Running time: 90 minutes
Historical context: Korea

Tagline: Of all the screen stories of young men in action, this is the most moving, the most honest, the most memorable!

Cast

Alan Ladd - Sergeant Kincaid
Sidney Poitier - Sergeant Eddie Towler
James Darren - Private Cotton
Glenn Corbett - Private Wade
Mort Sahl - Corporal Crane
Ingemar Johansson - Private Torgil

Quote: "Spit out what's on your filthy little mind, and then take your orders from me!" – Sergeant Towler

Highlights

This was one of the first films to take a serious look at racism by exploring the racial integration of the American military, and centers on an African-American sergeant's struggle to win the trust and respect of the men in his unit. It is a great example of how the Marine Corps was way ahead of the general population in the area of race relations as a direct result of the instinct to survive.

Reel Marines

Plot

In October of 1950, shortly after U.S. forces invade Korea, an advance Marine unit is sent to find and hold a farmhouse situated in a strategic mountain pass, and as the Marines make their way down a snow-covered mountainside they are attacked by waiting Chinese troops. The patrol leader is mortally wounded, and just before he dies Lieutenant Toland orders Sergeant Eddie Towler, the unit's only black member, to take charge of the few surviving Marines even though Towler suggests that Sergeant Kincaid - a veteran who has been with the outfit for eleven years - is better prepared to direct the unit.

Towler guides the men across the slippery, heavily mined slopes, and during the trek Kincaid rescues one of the Marines when he slips and lands among the mines. Then as they reach the farmhouse one of the men panics and throws a grenade inside the courtyard walls, seriously injuring a Korean woman who lives there with her part-French daughter and grandson.

Once inside, the men begin to worry that the numerous Chinese troops in the area will kill them all before the advancing Marine battalion can reach them, but Towler orders them to hold their position at all costs. Bracken, a Southern bigot, claims black men are unsuited to be leaders, and Kincaid thinks the men should be moved even though it would mean losing the pass to the enemy and endangering their entire battalion, and even suggests that Towler wants to remain in the farmhouse merely to prove himself. Determined to carry out Toland's dying command, Towler claims he will kill any man who refuses to act like a Marine and defend the farmhouse.

Reel Marines

That night, as Towler contemplates their difficult situation, the men reminisce about home while a recent immigrant from Sweden named Torgil, who wants to become a citizen and bring his family to the U.S., sings a song from his native land. Crane, a cynical corporal, tells amusingly irreverent stories about high-ranking officers, and a young Marine named Cotton sings and accompanies himself on a Korean stringed instrument.

Before long, an enemy patrol unit advances on the house and the shooting begins. The Marines repulse the attack and Hunter, a Navajo from Arizona nicknamed "Chief," volunteers to scout the area for other enemy soldiers. The Chinese capture Hunter and bring him back to the farmhouse, but in order to save his unit he refuses to give the password and dies with his Chinese captors when Towler and Kincaid fire on the intruders. After Hunter's burial the Eurasian woman living in the house thanks Towler for helping her, and tells him that one day his color will make no difference to the others. After another battle with the Chinese which costs the life of one of the men Towler and Kincaid come to blows, but their fight is interrupted by the sound of an approaching tank. Together they sneak onto the tank and set it ablaze, but Kincaid's leg is crushed as he tries to get out of the way and it must be amputated by a Corpsman who is unsure of his ability. Encouraged by Towler, the Doc continues with the operation, and Towler donates his own blood to keep Kincaid alive despite Bracken's objections. As a line of Chinese tanks approaches Towler orders the Marines to safer ground and protects Kincaid during an explosion, and then carries him out of the farmhouse. The enemy is about to reach the two men when U.S. planes appear overhead and blast the Chinese troops, and in the end Towler and Kincaid wish each other a Merry Christmas.

Reel Marines

Trivia

- The Marine Corps no longer had separate black units and bases after 18 November, 1949.

- Hall Bartlett designed a film for Sidney Poitier about the integration of the military during the Korean War, but only Columbia Pictures would finance it, and they insisted he rewrite the script to include a white co-star. Bartlett found the only major star willing to do the film was Alan Ladd, who also co-produced.

- Alan Ladd's daughter-in-law, Cheryl Ladd, would star in another movie which is featured in this book, *Purple Hearts*, in 1984.

- The film has an unusual cast which works well together. In addition to Alan Ladd and Sidney Poitier, the cast features Mort Sahl doing a comedy routine, James Darren singing the title song, Paul Richards as a bigoted Southerner, and boxer Ingemar Johansson in his American film debut.

- Hall Bartlett cast his Argentine wife Ana María Lynch (credited as Ana St. Clair) as a Korean.

- All the Young Men was filmed in Glacier National Park in Montana and on Mt. Hood in Oregon.

- Local Blackfoot Indians were cast as North Koreans.

- Columbia planned two separate advertising campaigns for the film, one each for white and black audiences. Columbia also used noted war correspondent Quentin Reynolds to promote the film in advertising campaigns.

Reel Marines

❖ A paperback novelization of the film was written by Marvin Albert.

❖ The Marine Corps provided Lieutenant Colonel Clement J. Stadler, who had been awarded the Navy Cross in World War II, as a technical advisor - a function he also performed in *Hell to Eternity, The Outsider, Ambush Bay* and for *The Lieutenant* television series.

❖ Swede Ingemar Johansson, who played Private Torgil, was the reigning Heavyweight Champion of the World during filming.

Reel Marines

> Release date: 14 September 1966
> Running time: 109 minutes
> Historical context: World War II
>
> Tagline: Their top secret mission paved the way for the man who said, "I Shall Return!"
>
> **Cast**
>
> Hugh O'Brian - First Sergeant Steve Corey
> Mickey Rooney - Gunnery Sergeant Ernest Wartell
> James Mitchum - PFC James Grenier
> Peter Masterson - Platoon Sergeant William Maccone
>
> Quote: "Stay away from things you've never done... like thinking." - 1stSgt Corey to PFC Grenier

Highlights

This movie is notable for the performances of Marine Corps veteran Hugh O'Brian, who is recognized as the youngest Drill Instructor to have served in the Corps, and Mickey Rooney, whose invitation for his Japanese captors to join him in having "pineapples" for lunch is comical. The film's underlying plot also offers a tangible, although fictional, explanation for the old ditty, "With the help of God and a few Marines, MacArthur returned to the Philippines!"

Reel Marines

Plot

Prior to the 1944 American invasion of the Philippines a hand-picked team of United States Marine Corps amphibious reconnaissance scouts is landed by PBY Catalina with the mission of contacting an intelligence agent who has crucial information. Each Marine is not only experienced, but has a special skill - with the exception of the radio operator, PFC Grenier.

Grenier, an air crew radioman with only six months in the Corps, is taken off the PBY's air crew when the original radio operator suddenly became medically unfit for the mission. He is given the sick Marine's radio and camouflage jacket to carry on his first ground combat mission, and also serves as the film's narrator.

After meeting up with their guide (Manual Amado) the patrol commander, Captain Alonzo Davis (Lieutenant Colonel Clement Stadler, who also acted as the film's technical advisor) is killed while ambushing a small group of Japanese soldiers, and First Sergeant Corey takes command.

Private George George (that's not a misprint) and PFC Henry Reynolds are killed while taking out a Japanese tank and patrol, and Corporal Stanley Parrish is killed by a guerilla trap soon after. As they walk on Amado is shot by a Japanese officer while scaling a small hill, and the Marines let him die to keep their presence a secret. Grenier is eventually told by Gunnery Sergeant Wartell that they were sent to recover some important information from a contact in a tea house whose radio was destroyed, thus explaining the radio operator's importance to the mission.

Grenier's inexperience and incompetence arouses the anger of Corey and the other members of the patrol, and his only friend is the easygoing but professional Gunnery

Reel Marines

Sergeant Wartell, who acts as a mediator between the no-nonsense Corey and Grenier, and explains each to the other as well as the audience. The surviving squad members eventually arrive at the tea house, but unfortunately Amado was the one who was supposed to meet the contact as he was the only Filipino in the group. Desperate, the group sends Corey to meet the contact, a Filipino-American woman from Long Beach, California named Miyazaki. While sneaking out of the camp with Miyazaki Corey crashes into a waiter, and the two run across a straw bridge before blowing it up with a grenade to escape from pursuing soldiers. Meanwhile a large skirmish with a Japanese patrol has killed Corporal Alvin Ross and Platoon Sergeant William Maccone, shot up the radio beyond repair, and wounded Gunnery Sergeant Wartell. Wartell, knowing he will slow the survivors down, tells the reluctant Corey to leave him behind, and when they leave he plants grenades under himself. He is soon captured by Japanese soldiers, and after toying with them for a bit during his interrogation he sets off the grenades and takes them all out in the blast - leaving Corey and Grenier as the only surviving Marines. The explosion is heard by the survivors, and they sadly move on.

Corey and Grenier learn fom Miyazaki that the Japanese are expecting the invasion and have placed a mine field powerful enough to destroy the entire fleet in the water around the expected invasion sites. Arriving at a friendly Filipino village, Corey and Grenier are able to escape a Japanese patrol by boat but Miyazaki is killed by a officer she seduced to buy them time.

Grenier, who has at last discovered the principles of mission accomplishment, altruism, and self-sacrifice through his observation of the others, becomes a squared away

Reel Marines

Marine. He and his First Sergeant infiltrate the enemy base to destroy the minefield, and plan to detonate it with a Japanese radio transmitter. As Corey provides a one-man diversion, Grenier is able to detonate the mines by radio control. Grenier then steals a radio and goes to tell Corey of their success, only to find him dead of blood loss from wounds sustained while holding off the Japanese. Grenier escapes to the coast and radios for pick up, the sole survivor of the mission, and the movie ends with him looking at the ocean as he listens to General Macarthur's speech.

Trivia

❖ Marine Lieutenant Colonel Clement J. Stadler acted as a technical advisor for this film as he had for *All the Young Men, Hell to Eternity, The Outsider*, and *The Lieutenant* television series.

❖ A novelization of the film's screenplay was written by Jack Pearl.

❖ During filming Mickey Rooney became ill with a fever, and while he was hospitalized in Manila his wife Barbara Ann Thomason was killed by her lover in a murder/suicide.

❖ Because the movie was shot on location in the Philippines on a modest budget, Filipino actors play the Japanese soldiers in the film.

❖ James Mitchum is the oldest son of Robert Mitchum, who has also played Marine roles in *Gung Ho!* and *Heaven Knows, Mr. Allison*.

Reel Marines

AN OFFICER AND A GENTLEMAN

Release date: 28 July 1982
Running time: 122 minutes
Historical context: Peacetime

Tagline: Life gave him nothing, except the courage to win... and a woman to love.

Cast

Richard Gere - Officer Candidate Zack Mayo
Louis Gossett, Jr. - Gunnery Sergeant Emil Foley
Debra Winger - Paula Pokrifki
David Keith - Officer Candidate Sid Worley
David Caruso - Officer Candidate Topper Daniels
Robert Loggia - Chief Petty Officer Byron Mayo

Quote: "The only two things from Oklahoma are steers and queers, and I don't see any horns on you boy." - GySgt Foley

Highlights

The appeal of this movie begins and ends with Louis Gossett Jr.'s outstanding portrayal of Gunnery Sergeant Foley. The difference between Foley and sailors like Chief Mayo would be comical if their characters weren't so authentic, and "old salts" who have spent time in the Philippines will also enjoy the brief sequences filmed there for flashbacks to Zack's youth. "You cherry boy!"

Reel Marines

Plot

An Officer and a Gentleman is a drama which tells the story of a U.S. Navy aviation officer candidate who comes into conflict with the Marine Corps Gunnery Sergeant who trains him.

The film begins with Zachary "Zack" Mayo receiving a college graduation present from his father Byron, a brash, womanizing U.S. Navy Boatswain's Mate formerly stationed at Subic Bay in the Philippines, where Mayo moved to live in early adolescence after his mother commits suicide. Aloof and taciturn with repressed anger at his mother's death and his father's inability to properly parent him, Mayo surprises his father when he announces his aspiration to be a Navy pilot.

Once he arrives at the thirteen-week long Aviation Officer Candidate School (AOCS), Mayo runs afoul of his abrasive, no-nonsense drill instructor, Marine Gunnery Sergeant Emil Foley. Mayo, or "Mayonnaise" as he is dubbed by the irascible Foley, is an excellent officer candidate, but not a team player. Foley rides Mayo mercilessly, sensing the young man would be prime officer material if he were not so self-involved. Zack becomes friends with fellow trainee Sid Worley, who is from the "good side of the tracks," and whose father and late brother were also Naval officers. Another focus is female recruit Casey Seeger (Lisa Eilbacher), whose name is pronounced like Bob Seger, but whom Foley calls "Cigar." She is unable to get over a wall with a rope on the obstacle course, and endures her own barrage of pressure from Foley.

Zack and Sid meet two local girls at a Navy-hosted dance, factory workers Paula Pokrifki and Lynette Pomeroy (Lisa Blount), who bed the cocky officer candidates. Foley has

Reel Marines

warned the officer candidates about the local girls (the "Puget Sound Debs") who look upon the "OCs" as potential husbands in order to escape their lower middle class lifestyles. Lynette appears to be the quintessential "Deb" who is trying to nab an officer candidate so she can discard her drab, blue-collar life and become an "aviator's wife." Sid takes up with Lynette eagerly and naively, but views his relationship with her as little more than sexual recreation. Paula is different. She makes no demands, and is content to let the relationship with Zack be what it is.

When Mayo's side business of selling pre-shined belt buckles and shoes to his fellow OCs is discovered by Foley, the drill instructor makes life unendurable for the trainee in order to force his resignation from the program. He asks for his "Drop On Request" (DOR), but Mayo refuses to give in. When Foley finally threatens to simply discharge Mayo, Zack breaks down and admits, "I got nowhere else to go! I got nowhere else to go... I got nothin' else." Satisfied that Mayo has come to a crucial self-realization, Foley lets up on him as Mayo begins to mature and mend his ways.

During a night of passion, Mayo reveals to Paula the truth behind his mother's suicide and that he is seeking a different path from that of his father. Paula later takes Zack home to "meet the family," and he learns Paula's biological father was in fact an officer candidate who refused to marry her mother when she was pregnant with her.

Later, Mayo is running with Seeger through the obstacle course one last time. Zack has a chance to break the record time for negotiating the course, but after Seeger fails once again to get over the wall he chooses to sacrifice the record to encourage her so she can graduate, which becomes a defining moment in Mayo's transformation to being a "team player."

Reel Marines

    As graduation nears Zack begins to distance himself from Paula, and at a dinner with Sid and his parents in town learns Sid has a long-time girlfriend back home whom he plans to marry shortly after commissioning - although he intends to continue his sexual relationship with Lynette until graduation. Meanwhile Lynette begins dropping hints to Sid that she might be pregnant, which adds to the pressure he is already experiencing in the AOCS program. During a high-altitude simulation in a pressure chamber he has a sudden anxiety attack and is attended by Zack, who tries to calm him down under the watchful eye of Foley. Realizing he joined the AOCS program because of expectations from his family and a sense of obligation to his brother, Sid DOR's without telling Zack. Believing that Foley pressured Sid to do it, Zack confronts Sid and Foley to try to get his friend reinstated and in doing so argues passionately that Sid is an ideal officer candidate only to find out that Foley also tried to talk Sid out of it – so the decision was completely Sid's.

    When Sid subsequently proposes to Lynette she turns him down, but not before confessing she wasn't pregnant as they originally thought. She wanted him to graduate in order to fulfill her dream of marrying a Naval aviator, and all but curses him for dropping out in the twelfth week. Despondent, Sid later commits suicide, and Mayo unreasonably blames Foley. The two clash in an unofficial, no-holds-barred martial arts bout with several of the candidates looking on. While Mayo physically dominates most of the match due to his youth, anger, and prior training, Foley finally wins after he kicks Mayo in the groin to maintain his authority over Mayo and all the other candidates. With both of them hurt, Foley offers him the chance to DOR one last time, knowing by now Mayo has either burned out his rebelliousness and misplaced anger, or never will.

Reel Marines

Mayo graduates with the rest of his class, and following the tradition of newly-commissioned Naval officers he seeks out and receives his first salute from Foley in exchange for a silver dollar. Mayo then thanks Foley, saying he'll never forget him, and Foley, clearly moved, responds with "I know" before straightening and giving Mayo a picture-perfect salute. Shortly thereafter Mayo rides from base on his motorcycle and stops to hear Foley verbally dressing-down the newest batch of AOCS recruits. He smiles, because he knows the process is beginning all over again.

In the iconic final scene of the film newly commissioned Ensign Mayo goes to the factory where Paula works wearing his Dress Whites, picks her up, and walks out carrying her in his arms. Lynette watches bitterly at first, knowing her own manipulations have left her alone in the end, but then applauds her friend along with the rest of the factory workers and shouts, "Way to go, Paula!"

Trivia

- ❖ The film's title is derived from an old expression from the British Royal Navy and the U.S. Uniform Code of Military Justice, as being charged with "conduct unbecoming an officer and a gentleman" (circa 1860).

- ❖ The film was shot in late 1981 on the Olympic Peninsula of Washington State at Port Townsend and Fort Worden because the Navy did not permit filming at NAS Pensacola in western Florida, the traditional site of the Aviation Officer Candidate School. Some early scenes were also filmed in Bremerton, with ships of the Puget Sound Naval Shipyard in the background.

Reel Marines

- A real motel, "The Tides Inn," located in Port Townsend was used in the film. Today there is a plaque outside the room commemorating this appearance.

- The "Dilbert Dunker" scenes were filmed in the swimming pool at what is now Mountain View Elementary School (Mountain View Middle School during filming). The dunking machine was constructed specifically for the film, and was an exact duplicate of the one used by the Navy.

- The concrete structure where Richard Gere delivered the line, "I got no place else to go!" is Battery Kinzie, which is located at Fort Worden State Park. The obstacle course was constructed specifically for the film, and was located in the grassy areas just to its south.

- The decompression chamber was one of the only sets constructed for the film and as of 2009 it is still intact in the basement of building number 225 of Fort Worden State Park. It can be seen through the windows of the building's basement.

- The blimp hangar used for the fight scene between Louis Gossett Jr. and Richard Gere is located at Fort Worden State Park and as of 2009 it is still there, but has been converted into a 1200 seat performing arts center.

- Director Taylor Hackford kept Lou Gossett Jr. in separate living quarters from the other actors during filming so he could better intimidate them during his scenes as a drill instructor.

- Gossett was advised by Marine Corps Gunnery Sergeant Buck Welcher.

Reel Marines

- John Travolta turned down the lead role (which went to Richard Gere) on the advice of his agent. Originally, country music singer John Denver signed on to play Zack Mayo, but he later opted out, saying it read like a "1950's movie." The casting process also included Jeff Bridges and Christopher Reeve.

- The role of Paula was originally given to Sigourney Weaver, then to Anjelica Huston and later to Jennifer Jason Leigh, who dropped out to do *Fast Times at Ridgemont High* instead. Kristy McNichol and Brooke Shields were also offered the role of Paula, but both turned it down.

- Rebecca de Mornay, Meg Ryan, and Geena Davis, all virtually unknown at the time, also auditioned for the role of Paula before losing out to Debra Winger.

- Jack Nicholson turned down the role of Gunny Foley, and no one else the producers were interested in was available. Screenwriter Stewart then visited the Pensacola area to do research and found out that all of the top drill instructors there were African-American, which inspired them to cast Louis Gossett Jr. in the role for which he won an Oscar.

- Lisa Eilbacher, who played Officer Candidate Casey Seeger, is an avid bodybuilder/fitness buff and said pretending to be out of shape for her role was the most difficult part about acting in the film.

- Richard Gere rides a 750cc T140E Triumph Bonneville in the movie.

Reel Marines

- Producer Don Simpson unsuccessfully demanded the ballad *Up Where We Belong* be cut from the film, saying, "The song is no good. It isn't a hit." The song later became number one on the Billboard chart and won the Academy Award for Best Song. Simpson wanted a similar song called *On the Wings of Love* by Jeffrey Osborne, which was released a few months later and peaked at number twenty-nine on the Billboard charts.

- It is a Naval tradition for newly-commissioned officers to give a silver dollar to the person who gives them their first salute. In the scene where the graduates of Foley's class receive theirs, you can see them giving him a silver dollar prior to each salute. It is also a tradition for the D.I. to place the silver dollar of his memorable students in his right pocket, as he does with Mayo's.

- At the graduation ceremony when Zack says he's going to get his first salute he is referring to his father, a reference to the opening scenes when Zack's Dad said he'd never salute him. A scene was shot at the graduation where Zack's father salutes him, but it wasn't used. Robert Loggia protested it being cut, and the footage is considered lost.

- Debra Winger negotiated her own contract before she saw the revised script and was not happy when she found out she would be doing a nude scene. She asked to be covered up, but was told that since she hadn't thought to ask for a "no nudity clause" in her contract she would have to do it as written.

Reel Marines

BATTLE CRY

Release date: 2 February 1955
Running time: 149 minutes
Historical context: World War II

Tagline: The men who fought, the women who waited, and the stolen moments they shared.

Cast

Van Heflin - Major Sam Huxley
James Whitmore - "Mac"
Tab Hunter - PFC Danny Forrester
Aldo Ray - PFC Andy Hookens

Quote: "The call me Mac. The name's unimportant. You can best identify me by the six chevrons, three up and three down, and by that row of hashmarks. Thirty years in the United States Marine Corps." - Mac

Highlights

In addition to being an enjoyable World War II era film, Battle Cry is significant because it was written by a Marine (Leon Uris) and stars a Marine (James Whitmore). The scene which best represents the "Marine mentality" is the one in which Colonel Sam Huxley risks court-martial and demands a MORE dangerous assignment for his battalion in the Saipan landing. I would not expect anyone who has not been a Marine to understand!

Reel Marines

Plot

In January of 1942, as many young men respond to the call for Marine Corp recruits, All-American athlete Danny Forrester boards a train in Baltimore, Maryland after saying goodbye to his family and girlfriend Kathy. The train picks up other recruits enroute to the Marine training camp near San Diego, including womanizing lumberjack Andy Hookans, bookish Marion Hodgkiss, Navajo Indian Shining Lighttower, troublemaking "Spanish" Joe Gomez, L. Q. Jones of Arkansas, Speedy of Texas, and the Philadelphian Ski, who is eager to escape the slums, but upset to leave his girlfriend Susan.

Several weeks later, after the arduous training of boot camp, the men are accepted into radio school and assigned to the battalion commanded by Major Sam "High Pockets" Huxley. The Marines continue their military training and receive rigorous communication instruction from "Mac," the grizzled Master Sergeant who heads their section, but on weekends they get passes to San Diego. In a sleazy bar there, Ski drowns his sorrows in alcohol and women to forget that his Susan has married another man. Concerned about him, Mac and his fellow Marines go to the bar, believing they are coming to his rescue, and get into a brawl. Danny is saved from excessive drinking by married USO worker Elaine Yarborough and begins a relationship with her until Mac, noticing a change in his performance, arranges for him to call Kathy long-distance. Recognizing the young man's loneliness, Mac and Huxley grant him a furlough to Baltimore during which Danny elopes with Kathy. Meanwhile the meditative Marion, who hopes to write about his wartime experiences, meets the beautiful and mysterious Rae on the Coronado ferryboat. Although she meets him there frequently and seems to admire him greatly she will not share with him details about her life, and Marion learns why she has been evasive when she shows up with other B-girls at a party celebrating the regiment's orders to ship out.

The men are sent to Wellington, New Zealand, where they are warmly received. Andy, who respects no woman, tries to woo the married Pat Rogers by suggesting he fill the void left by her husband, whom he believes is fighting in Africa. After the offended Pat tells him her husband died in action, Andy apologizes for the first time ever. Pat later invites the reformed Andy to visit her parents' farm where, despite their attraction, they agree to remain friends only. After Christmas the Sixth Regiment, now known as "Huxley's Harlots," is sent to Guadalcanal after the invasion to "mop up" a resistant band of Japanese soldiers. Afterward the battle-weary men, minus Ski, who was killed by a sniper, return to New Zealand where Pat nurses the malaria-stricken Andy and decides to risk a short-term romance with him.

To restore the men's stamina Huxley, newly promoted to lieutenant colonel, orders them to complete a brutal sixty-mile hike, and while other units are trucked back to camp Huxley has his men hike the whole way, blistered and near collapse, but in record-breaking time.

Aware that his men are special, Huxley is frustrated when they are not ordered to Tarawa with the main invasion, but instead are held back to clear out remaining Japanese resistance afterward. Pat is afraid of losing another love to the war and tells Andy that she wants to break up, but Andy refuses and asks her to marry him. Although frightened, she accepts, and only then admits that she is pregnant. With Huxley's assistance in cutting through red tape Andy and Pat marry, but two days later when the men are to ship out Andy

Reel Marines

considers deserting to stay with Pat. Instead of arresting him, Huxley asks Pat to convince Andy to return voluntarily. At Tarawa Huxley's men fulfill their mission, but Marion and many others are killed. Afterward, while standing by in reserve on a Hawaiian island, Huxley receives word that other battalions are being moved out for the invasion of Saipan. Sensing the restlessness of his men, Huxley risks court-martial to convince General Snipes that the talents of his battalion are being wasted. Although at first offended by Huxley's "impudence," Snipes assigns the battalion to the invasion of Red Beach, the most dangerous mission in the Saipan campaign.

After the landing the men are isolated from the rest of the division, and suffer heavy casualties from artillery fired from the hills above them. Huxley is killed, and Danny and Andy are seriously wounded, but the battalion holds out until a Navy destroyer pins down the Japanese and frees the Marines to complete their mission.

Later, at a rest camp recuperating from the loss of a leg, Andy becomes too demoralized to communicate with Pat or his concerned friends, but tough words from Mac make him realize that Pat still loves him. Andy then returns to her and his baby son after completing rehabilitation.

Danny is also given a medical discharge and returns by train to Baltimore, accompanied by Mac, who is visiting the families of men killed in action and seems to be alone as the scene opens, leading the viewer to initially believe Danny has been killed. In Baltimore they say goodbye, and Danny reunites with the waiting Kathy as fresh recruits board the train.

Reel Marines

Trivia

- *Battle Cry* was adapted for the screen from a 1953 novel by Leon Uris. Many of the events in the book are based on Uris' own World War II experiences with the 6th Marine Regiment.

- Uris joined the Marine Corps at the age of seventeen and served in the South Pacific at Guadalcanal, Tarawa, and New Zealand from 1942 to 1945.

- Leon Uris also wrote the screenplay.

- The film was produced and directed by Raoul Walsh, who had also directed the original (silent) version of *What Price Glory?* in 1926.

- Battle Cry received an Academy Award nomination for Scoring of a Dramatic or Comedy Picture.

- This was the film debut of L.Q. Jones (billed under his real name of "Justus E. McQueen"), who took his character's name in *Battle Cry* as his legal name after the film was released.

- L. Q. Jones became one of famed director and former Marine Sam Peckinpaugh's favorite actors, and appeared in several of his films.

- James Whitmore, who played Mac, served in the Marine Corps during World War II.

Reel Marines

BEACHHEAD

Release date: 5 February 1954
Running time: 90 minutes
Historical context: World War II

Tagline: It's not just a war movie.

Cast

Tony Curtis - Burke
Frank Lovejoy - Sergeant Fletcher
Skip Homeier - Reynolds
Alan Wells - Biggerman
Eduard Franz - Bouchard
Mary Murphy - Nina Bouchard

Quote: "Fight your own battles inside and out, but don't bring me personal troubles when I give you an order." - Sgt Fletcher

Highlights

The star of Beachhead is not Tony Curtis in the first of his two roles as a Marine, beautiful Mary Murphy as the token female, or even Frank Lovejoy as the tough career Marine leading the mission. The star is the jungle. If you have ever spent time training or fighting in the jungles of Central America or Southeast Asia you will be reminded of what it was like, and if you have never been there you will develop an appreciation for that nasty, unforgiving environment.

Reel Marines

Plot

During the World War II Western Pacific campaign the island of Bougainville is occupied by the Japanese. Thirty miles away Sergeant Fletcher, a twenty-six-year Marine Corps veteran, is informed by Major Scott that a full-scale invasion of Bougainville is planned. Scott also states that a French plantation owner named Bouchard has radioed information about the location of Japanese minefields on the beaches, and assigns Fletcher to find the Frenchman and verify the information in order to save lives during the invasion.

Fletcher, who is guilt-ridden about having lost most of his platoon at Guadalcanal, sets out with his remaining men - Burke, Reynolds and Biggerman - to locate Bouchard. While following a river they come upon a dozen Japanese soldiers emerging from a swim, and although it is not part of their mission to engage the enemy they shoot a number of them - and lose Biggerman when he attempts to drop a grenade into a Japanese tank. Before nightfall the three men dig a trench, hoping to ambush the remaining Japanese soldiers who are following them, however when a flare reveals their position and they come under fire Burke crawls out of the foxhole with his knife drawn and struggles with a sniper in the undergrowth. Reynolds goes to his assistance but is killed, and after Fletcher and Burke bury him they move on.

Eventually they reach Bouchard's plantation where they discover his daughter Nina, who then takes them to where her father is observing enemy activities. After Bouchard shows the Marines a map with the safe channel clearly marked Fletcher and Burke decide to radio Scott with the confirmation, so they head for an enemy radio hut which Bouchard has used as it is normally lightly guarded. They are followed by the Japanese soldiers however, and after discovering that the hut and surrounding area are booby-trapped the group manages to lure their pursuers into the compound and blow it up. The four retreat into the jungle and decide to head for their rendezvous point so that they can deliver the information in person.

Meanwhile Burke is unhappy that Nina has become part of their problem, and Fletcher sends him and Bouchard to scout a trail. Nina is aware of the animosity between Burke and Fletcher over the deaths of so many men, and when she is alone with Fletcher he admits to her that he feels he has always made the wrong decisions. A short time later Burke and Bouchard return with a Japanese deserter who appears very frightened when approached by a native carrying a machete. Bouchard talks with the native, and he agrees to take them where they want to go in exchange for the deserter. After the native leads them to a riverbank and a canoe Fletcher leaves the deserter with him, and as they pull away cries are heard.

The river journey is successful until they hear a patrol boat and are forced to pull onto shore and abandon the canoe. Nina falls and injures her ankle, but they proceed through the jungle on foot while being followed by Japanese from the patrol boat. Because Bouchard says the rendezvous spot is still many hours away and Nina is having difficulty walking on her swollen ankle it is unlikely they will arrive in time, so Fletcher suggests that Burke go on alone. Burke does not want to leave Nina, to whom he has become romantically attached, and slugs Fletcher.

Suddenly a shot rings out and Bouchard drops dead, the victim of a sniper. Fletcher then sends Burke and Nina on ahead, draws the sniper's fire, and eventually kills him. Burke and Nina struggle on, but she asks him to leave her in

Reel Marines

order to get the map through and save many lives in the coming invasion. As they rest Fletcher catches up with them and angrily forces them on until they come to a ridge from which they can see the pier and the waiting boat. As they advance along the beach the pier is fired on by a Japanese battleship, and an American patrol boat responds but is blown out of the water. Burke then swims out with a grenade and ignites the river of gasoline which extends from the wreckage to the battleship, causing it to explode. He then swims back to shore, but their boat has left without them.

The three of them sit on the pier expecting the Japanese eventually to get them, but instead a squad of Marines led by Major Scott arrives. After Fletcher and Burke hand over the map another boat pulls in and Nina is evacuated, but not before Burke tells her that he will be seeing her. Burke and Fletcher, who now respect each other, rejoin their unit.

Trivia

- *Beachhead* is based on Marine Captain Richard G. Hubler's 1945 novel *I've Got Mine*.

- The movie was filmed on Kauai in the Hawaiian Islands.

- The producers used Hawaiians for many of the roles including Sam "Steamboat" Mokuahi, Democratic Party organizer Dan Aoki, and Akira Fukunaga. The latter two are veterans of the 442nd Regimental Combat Team.

- The producers went to the Marine Corps to seek technical assistance for the making of the film, and although the Corps liked the idea they refused to provide cooperation because two of the four Marines were killed in the screenplay. The Public Information Officer said the Marines would not provide assistance to any film

Reel Marines

showing the Corps taking fifty percent casualties as they were in the midst of a recruiting campaign.

❖ The producers visited the Pentagon and were provided Navy, Coast Guard, and Hawaiian National Guard assistance in making the film.

❖ The film was titled *Missione Suicido* (*Suicide Mission*) in Italy, which served to bear out the Marines' point.

❖ Mary Murphy, who played Nina Bouchard, felt director Stuart Heisler was trying to make her look like a version of his own wife.

❖ Mary Murphy was nearly attacked by a drunken cameraman in her room at the film's isolated Hawaiian location, but was rescued by stars Frank Lovejoy and Tony Curtis.

❖ Tony Curtis also starred as a Marine in *The Outsider*, playing Iwo Jima flag raiser Ira Hayes.

Reel Marines

BORN ON THE FOURTH OF JULY

Release date: 20 December 1989
Running time: 145 minutes
Historical context: Vietnam

Tagline: A story of innocence lost and courage found.

Cast

Tom Cruise - Ron Kovic
Tom Berenger – Gunnery Sergeant Hayes
Willem Dafoe – Charlie
Stephen Baldwin - Billy Vorsovich
Kyra Sedgwick – Donna

Quote: "Don't you know what it means to me to be a Marine, Dad?" - Ron Kovic

Highlights

While Born on the Fourth of July is not pretty to watch, it serves as a reminder of what happens when a Nation forgets its veterans. Some people believe the movie is an indictment of the military, but it is not. Kovic is proud to be a Marine, and he and his brothers in arms fight valiantly. The villains in this film are politicians, bureaucrats, and people who are unable to separate the policies of a conflict from the men who are fighting it. While still imperfect, it seems the American people have taken this message to heart - if the treatment if Iraq veterans is any indication.

Reel Marines

Plot

After Ron Kovic and his fellow students hear an impassioned lecture about the Marine Corps, Ron decides to enlist. He misses his prom because he is unable to secure a date with his love interest, Donna, but he confronts her there and they dance on his last night before leaving.

The film then fast forwards to Kovic's second Vietnam tour in 1968. Now a Marine sergeant and on patrol, his unit massacres a village of Vietnamese citizens, believing them to be enemy combatants. During the withdrawal Kovic becomes disoriented and accidentally shoots one of the new arrivals to his platoon, a younger Marine private first class named Wilson. Despite the frantic efforts of the Navy Corpsman to save him, Wilson later dies from his wounds and leaves a deep impression on Kovic. Overwhelmed by guilt, he appeals to his executive officer, who merely tells him to forget the incident. The meeting has a negative effect on Ron, who is crushed at being brushed off by his XO.

The platoon goes out on another hazardous patrol a few weeks later. During a firefight Kovic is critically wounded and trapped in a field facing sure death until a fellow Marine rescues him. Paralyzed from the mid-chest down, he spends several months recovering at the Bronx Veterans Administration hospital. The hospital living conditions are deplorable. Rats crawl freely on the floors, the staff is generally apathetic to their patients' needs, doctors visit the patients infrequently, drug use is rampant (among both the staff and patients), and the equipment is too old and ill-maintained to be useful. Ron desperately tries to walk again with the use of crutches and braces despite repeated warnings from his doctors, and soon suffers a bad fall which causes a compound fracture of his thighbone. The injury nearly robs him of his leg, and he vehemently argues with doctors who briefly consider resorting to amputation.

Ron returns home, permanently in a wheelchair, but with his leg intact. While there he begins to alienate his family and friends with his complaints about students staging anti-war rallies across the country and burning the American flag. Though he tries to maintain his dignity as a Marine, Ron gradually begins to become disillusioned and feels his government has betrayed him and his fellow Vietnam Veterans. In Ron's absence his younger brother Tommy has already become staunchly anti-war, leading to a rift between them, and his highly religious mother also seems unable to deal with Ron's new attitude as a resentful, paralyzed veteran. His problems are as much psychological as they are physical, and he quickly becomes a belligerent alcoholic.

During an Independence Day parade Kovic shows signs of post-traumatic stress when firecrackers explode and a baby in the crowd starts crying. He reunites with his old high school friend Timmy Burns, who is also a wounded veteran, and the two spend Ron's birthday sharing war stories. Later Ron goes to visit Donna at her college in Syracuse, and the two reminisce. She asks him to attend a vigil for the victims of the Kent State shootings, however he cannot do so because his wheelchair prevents him from traversing the campus because of curbs and stairways. He and Donna are then separated when she and her fellow students are arrested for demonstrating against the war.

Ron's disillusionment grows severe enough that he has an intense fight with his mother when he returns home drunk one night after having a barroom confrontation with a World War II veteran who expresses no sympathy for him. He then travels to a small town in Mexico called "The Village of the Sun" which seems to be a haven for paralyzed Vietnam veterans. There he has his first sexual experience with a prostitute. He believes he's in love, and wants to ask her to marry him until he sees her with another customer. The understanding of real love versus a mere physical sexual experience sets in, and he decides against it. Ron then hooks up with another wheelchair-using veteran named Charlie, who is furious over a prostitute mocking his lack of sexual function due to his severe wounding in Vietnam, and the two travel to what they believe will be a friendlier village.

After annoying their taxicab driver the pair ends up stranded on the side of the road. They quarrel and fight, knocking each other out of their wheelchairs, and are eventually picked up by a man with a truck and driven back to the village.

On his way back to Long Island Ron makes a side trek to Georgia to visit the parents and family of Wilson, the Marine he believes he killed during his tour. He tells them the real story of how their son died and confesses his guilt. Wilson's widow, now the mother of the deceased Marine's toddler son, admits she cannot find it in her heart to forgive him for killing her husband, but Mr. and Mrs. Wilson are more forgiving and even sympathetic to his predicament and suffering because Wilson's father fought in the Pacific during World War II and is disillusioned with the war in Vietnam. In spite of the mixed reactions he receives, the confession seems to lift a heavy weight from Ron's conscience.

Ron next joins Vietnam Veterans Against the War (VVAW) and travels to the 1972 Republican National Convention in Miami where he and his compatriots force their way into the convention hall during Richard Nixon's acceptance speech and cause a commotion which makes it onto the national news. Ron tells a reporter about his

Reel Marines

negative experiences in Vietnam and the VA hospital conditions, and his interview is cut short when guards eject him and his fellow vets from the hall and attempt to turn them over to the police. They manage to break free from the police, regroup, and charge the hall again - though not so successfully this time. The film ends with Kovic speaking at the 1976 Democratic National Convention shortly after the publication of his autobiography *Born on the Fourth of July*.

Trivia

- Oliver Stone, himself a Vietnam veteran, read Ron Kovic's bestselling autobiography *Born on the Fourth of July* and was stunned by how Kovic had suffered after the war. He bought the rights to the book, and wanted to make it into a film.

- Kovic wrote the screenplay with Stone at a cafe in Venice, California and appears in the film as a WWII veteran who flinches at the sound of exploding firecrackers during the opening parade sequence - a reflex Cruise's Kovic adopts later in the film.

- Kovic and Stone received a Golden Globe for Best Screenplay and were nominated for an Academy Award, losing the Oscar to the screenplay of *Driving Miss Daisy* (which subsequently beat out *Born on the Fourth of July* as Best Picture).

- Stone wanted to film the movie in Vietnam, but since the relationship between Vietnam and the United States had not been resolved he decided to film it in the Philippines instead (where he had previously filmed *Platoon*).

Reel Marines

- *Born on the Fourth of July* is considered part of Oliver Stone's "trilogy" of films about the Vietnam War along with *Platoon* (1986) and *Heaven & Earth* (1993).

- Star Tom Cruise was actually born on the *third* of July.

- The entire film was shot in shades of red, white or blue, depending on the emotional level (battle scenes are in reddish hues, dream sequences in white, sadness in blue).

- Charlie Sheen was considered for the part of Ron Kovic.

- Abbie Hoffman did a cameo as a war/draft protester.

- Director Oliver Stone made a cameo appearance as the reporter interviewing a military official on television near the beginning of the film - played by the film's military technical advisor, retired Marine Captain Dale Dye.

- A copy of "Johnny Got His Gun," a popular anti-war novel about WWI, is visible on Ron's hospital bed when he learns he will never walk again. In that novel the main character is a soldier who is mutilated in the war, losing lost both of his arms and legs as well as his sight.

- Oliver Stone's then-wife Elizabeth is mentioned in the closing credit roll as Naijo No Ko. This Japanese term means "with the help of my wife," or more colloquially, "I owe my success to my better half."

- The word "fuck" is used 289 times throughout the film.

- For the duration of filming after Ron is paralyzed, Tom Cruise stayed in a wheelchair off-set as much as possible.

- The real Ron Kovic gave Tom Cruise his Bronze Star for his performance in this movie.

Reel Marines

DEATH BEFORE DISHONOR

Release date: 20 February 1987
Running time: 95 minutes
Historical context: Peacetime

Tagline: In a world of compromise, he wouldn't!

Cast

Fred Dryer - Gunnery Sergeant Burns
Brian Keith - Colonel Halloran
Paul Winfield - Ambassador
Joseph Gian - Sergeant Manuel Ramirez

Quote: "It's not just a job. It's a vendetta!" - Gunny Burns

Highlights

Death Before Dishonor is a great action movie if you don't mind a few "nuclear hand grenades." Recon Marines will find the infiltration exercise and "Gold Wing Ceremony" at the beginning interesting, and Embassy Marines will get a kick out of how they are portrayed as well. The film is notable for real-life Marine Brian Keith's role as Colonel Halloran, and for the scenes in which he and Sergeant Ramirez were tortured by Islamic terrorists. When you view their methods, which are not all that different in real life, it makes one wonder what all the fuss was about over Abu Ghraib and our use of water boarding!

Reel Marines

Plot

Gunnery Sergeant Burns is a Recon Marine, and as the movie opens he is training his troops in amphibious infiltration techniques and awarding gold parachute wings to those Marines who have earned them. He is subsequently assigned to take charge of the Marine Security Guard detachment at an American Embassy in a fictional Middle Eastern nation, and when terrorists attack the compound and take hostages Burns becomes a "one-man Marine Corps" in an attempt to rescue them and kill the terrorists.

Most of the problems facing Burns can be traced to the way the Ambassador, played by Paul Winfield, has tied the Marines' hands from an operational standpoint. In his "enlightened" view security must be sacrificed in order to avoid upsetting the host government, and his "rules of engagement" leave no margin for error.

Colonel Halloran, who was Gunny Burns' longtime Commanding Officer, is one the Americans who have been abducted by the terrorists, and he and Marine Sergeant Ramirez are mercilessly tortured by them with a power drill. The scenes in which this happens are graphic and extremely realistic, and serve as more than enough justification for retaliation to Burns' way of thinking - no matter what the Ambassador says.

Burns ignores the Ambassador's restrictions and throws the rule books out the window as he heads for a confrontation where he must rescue the hostages and wipe out the terrorists. His assault on their compound is bloody, and he takes out all of the opposition – including a pair of sadistic German terrorists from the Bader Meinhof Gang who have joined forces with the Islamo-fascists.

Reel Marines

Trivia

❖ In his 2001 book *Reel Bad Arabs*, Jack Shaheen cites *Death Before Dishonor* as one of the four most anti-Arab Hollywood movies ever. Just a few months later the World Trade Center was destroyed by Muslim terrorists on 9-11. Obviously, it wasn't as much of a sterotype as Mr. Shaheen would have had you believe.

❖ Brian Keith, who plays Colonel Halloran, served in combat as a Marine during World II. His biography is one of over a hundred in the book *Hollywood Marines*.

❖ Fred Dryer, who stars as Gunny Burns, played thirteen seasons in the NFL as a defensive end for the New York Giants and Los Angeles Rams before becoming an actor and starring in the 1984-1991 television police drama *Hunter*.

❖ At the beginning of the movie Gunny Burns is awarding "Blood Wings" to Marines who have just earned their Gold Navy/Marine Corps Parachutist Insignia. This involves placing the pins of the badge pointing into the chest of the jumper and then slamming them into his pectoral muscle. Once the pins have been driven into the flesh a punch, or pounding with a jump helmet, follows. Blood wings are against Armed Forces Policy and are prohibited, but recipients consider it a highly honorable rite of passage – although the real ceremony is not nearly as tame as the one depicted in the movie.

Reel Marines

EARS, OPEN. EYEBALLS, CLICK.

Release date: 2005
Running time: 155 minutes
Historical context: Peacetime

Tagline: This nonfiction portrait is an even purer distillation of famously brutal Marine training methods.

Cast

Platoon 1141, MCRD San Diego

Quote: "Answer my fucking question before I rip out your teeth!" - Drill Instructor

Highlights

When civilians watch movies like Full Metal Jacket, The Boys in Company C and The D.I. many of them think they are inventions of Hollywood – so a documentary like Ears Open, Eyeballs Click goes a long way towards dispelling that misconception. This film depicts the mundane parts of boot camp in addition to the high profile, high stress incidents which translate so well to feature films, so there is no doubt as to its authenticity. As is the case with the theatrical releases, touchy-feely liberals will tend to decry the Corps' "brutal" training methods while Marines who have actually gone through boot camp will begin their critiques with, "Back in the old Corps, it was harder..."

Reel Marines

Plot

There is no plot per se, but instead an observation of Platoon 1141, Company C, 1st Recruit Training Battalion, Marine Corps Recruit Depot San Diego, California from the initial bus ride to the depot through graduation. *Ears, Open. Eyeballs, Click* is 2005 documentary by Canaan Brumley about the experiences of Marine recruits during bootcamp. Unlike many documentaries this film offers no narration, nor does it focus on central characters. Instead it shoots from a fly-on-the-wall perspective. Despite this unusual approach the film has received very positive reviews overall, especially from film festivals such as the Los Angeles and San Diego Film Festivals.

The first moments of this documentary will bring Marines back to their first few hours of basic training with all of the confusion, fear and yelling. The Marine Corps obviously has the toughest boot camp of any branch of the military, but even so after the first few moments memories will fade and you will be able to follow the journey from an objective standpoint – and parents will be able to see firsthand what their sons go through to become a U.S. Marine. Poignant, funny, and frightening, this film is a must for anyone who has been in the service, is contemplating enlistment, or just wants to understand the journey to becoming a Marine.

Everything the DIs do is for a reason, although to the recruits, it doesn't seem like it. This documentary catches the day-to-day training and teaching which goes on, and shows the recruits at their worst, and at their best, as they try to navigate the course from civilian to Marine. It also captures the DIs in moments where they are "off," talking amongst themselves, as well as when they are "on," screaming to the point of incomprehension and trying to get a recruit to learn.

Reel Marines

Trivia

❖ Canaan Brumley began shooting the film with four cameramen, but only a few weeks into production they quit and he ended up shooting most of the footage himself.

❖ Brumley has handled distribution through DVD sales on his website. In addition, the film has been shown on the Documentary Channel.

❖ The film's title was derived from the fact that recruits in formation are prohibited from turning their head or eyes away from their direct front, even when being addressed. When a Drill Instructor speaks to a recruit, that recruit is expected to stare forward if the DI is oblique to him, and through him if the DI is directly to his front. When instructing recruits, a DI may command "Ears!" to which the proper response for the recruits is "Open, sir!" If commanding them to look at him or at something else, the DI may command "Eyeballs," to which the recruits also have a formulaic response, in this case "Click!"

❖ Brumley started directing films while attending high school in Houston and went on to receive his B.S. in film at San Diego State University. During this time he produced, wrote and directed thirteen short films before producing *Ears, Open. Eyeballs, Click,* which is his first feature.

❖ The Senior Drill Instructor for Platoon 1141 was Staff Sergeant Michael Nichols. *Ears, Open. Eyeballs, Click* was filmed during his first training cycle as a Senior.

Reel Marines

FIRST TO FIGHT

Release date: 21 May 1967
Running time: 92 minutes
Historical context: World War II

Tagline: The blockbusting story of a fighting Marine that comes mortar-screaming out of green hells and jungles!

Cast

Chad Everett – Sergeant Jack Connell
Dean Jagger - Lieutenant Colonel Baseman
Bobby Troup - Lieutenant Overman
Gene Hackman - Sergeant Tweed
Claude Akins - Captain Mason

Quote: "I just made a second lieutenant out of you. I know it's a come down from platoon sergeant, but we all have to make sacrifices." - Lieutenant Colonel Baseman

Highlights

First to Fight has a strong "Marine connection" in that writer Gene L. Coon and stars Gene Hackman and Bobby Troup all served in the Corps during World War II. The storyline of a hard charging combat Marine who becomes too civilized while "in the rear with the gear" is one to which many can relate, and serves as a cautionary tale for Marines who are tempted to live the good life and get soft rather than maintain their fighting edge.

Reel Marines

Plot

The Marines are dug in and ready in *First to Fight*, a World War II saga that takes its title from the *Marines' Hymn*, its battle savvy from the technical guidance of a Marine veteran of Saipan and Iwo Jima, and its power from a script by *Star Trek* veteran Gene L. Coon, himself a Marine sergeant during World War II and the Korean War.

Gunnery Sergeant "Shanghai" Jack Connell is the sole survivor of a Japanese attack on his squad at Guadalcanal, and because of his heroism and the fact that he is still alive he is promoted to second lieutenant and awarded the Medal of Honor – and yet his toughest test lies ahead.

After a stateside hitch during which he is decorated by the President, Connell is assigned to train new recruits for the Marines, and while on a war bond tour he falls in love with public relations woman Peggy Sanford. Peggy, whose fiancé was killed in the war, initially resists Jack's advances, but she agrees to marry him when he promises not to volunteer for combat duty.

When news comes that an old buddy has been killed on Bougainville, Jack begins to think of himself as a slacker and decides on a transfer back to the front lines. Sensing the change in her husband Peggy releases him from his promise, and he accepts orders to return to the Pacific... but Jack's no longer the gung-ho Devil Dog he used to be. Once back in combat Connell is frozen with fear under fire, and becomes too terrified to fight – until an unflinching Sergeant named Tweed (Gene Hackman) rekindles his courage. In the end Jack gradually overcomes his fear, and once more takes firm command of his men and leads them in a raid against a Japanese island stronghold.

Reel Marines

Trivia

- ❖ Gene Hackman, who plays Sergeant Tweed, left home at the age of sixteen to join the Marine Corps and served four-and-a-half years as a field radio operator.

- ❖ Hackman went on to portray another Marine, Colonel Cal Rhodes, in the 1983 film *Uncommon Valor.*

- ❖ The character of "Shanghai" Jack Conell is based on World War II Marine hero "Manila" John Basilone.

- ❖ Bobby Troup served as a Captain in the Marines during World War II, and was the first white officer to be given command of an all black unit.

- ❖ Troup's wife, Julie London, had previously been married to Jack Webb, who had starred as Marine T/Sgt Jim Moore in *The DI.*

- ❖ Bobby Troup wrote the music and lyrics for the song "Daddy" in 1941.

- ❖ Bobby Troup performed his own composition "Daddy" in the film.

- ❖ Star Chad Everett was chosen by the Wayne family to be the voice of John Wayne at Disney MGM Studios "Great Movie Ride."

- ❖ *First to Fight* was the first – and last – feature film in which female lead Marilyn Devin (Peggy Sanford) appeared.

Reel Marines

FLAGS OF OUR FATHERS

Release date: 20 October 2006
Running time: 127 minutes
Historical context: World War II

Tagline: The real heroes are the ones left on the island.

Cast

Ryan Phillippe - PhM2 John "Doc" Bradley
Adam Beach - PFC Ira Hayes
Barry Pepper - Sergeant Michael Strank
Jesse Bradford - PFC Rene Gagnon
Joseph Cross - PFC Franklin Sousley
Benjamin Walker - Corporal Harlon Block
Neal McDonough - Captain Dave Severance
Robert Patrick - Colonel Chandler Johnson
Chris Bauer - General Alexander Vandegrift

Quote: "People on the street corners, they looked at this picture and they took hope." – Bud Gerber

Highlights

Flags of Our Fathers is a special film because it treats the events surrounding the flag raising, and the flag raisers, as a historical event while retaining the drama and watchability of a theatrical movie. Great pains were taken to get the details right, and in doing so Clint Eastwood has done a great service to both the Nation and the Marine Corps.

Reel Marines

Plot

The story focuses on seven Marines of Easy Company, 2nd Battalion, 28th Marine Regiment, 5th Marine Division - Sergeant Mike Strank, PFC Rene Gagnon, PFC Ira Hayes, Corporal Harlon Block, PFC Franklin Sousley, Sergeant Hank Hansen, and PFC Ralph Ignatowski, as well as their Navy Corpsman, PhM2 John "Doc" Bradley.

In December of 1944 the Marines trained at Camp Tarawa, Hawaii by climbing a large mountain and getting into Higgins boats. They then set sail across the Pacific and it is revealed the task force is are headed to the small island of Iwo Jima, which is located seven hundred miles from the Japanese mainland. Captain Dave Severance tells them they will be fighting on Japanese soil, and should expect tough resistance. A few days later the armada arrives off the coast of Iwo Jima and the ships of the U.S. Navy open fire on suspected Japanese positions, and on the night before the landing Mike Strank is put in charge of second platoon.

The next day, February 19, 1945, the Marines hit the beach in landing craft and meet no resistance. Ralph, aka "Iggy," suspects that the Navy killed all the Japanese defenders, as do most of the Marines. After several tense minutes the Marines advance forward and the Japanese open fire. The battle is extremely intense, and the Marines take heavy casualties. Japanese heavy artillery opens fire upon the Marines ashore, as well as the Navy ships. After several attempts Second Platoon takes out a Japanese pillbox which was pinning them down and they advance forward, as do many other Marines.

The battle begins to calm down and the beachheads are secure, and two days later the Marines attack Mount Suribachi under a rain of Japanese artillery and machine gun

Reel Marines

fire as the Navy bombards the mountain. It is here that Doc saves the lives of several Marines under fire which later earns him the Navy Cross. Finally the mountain is secure, and for the next four nights the Marines take cover in foxholes as Japanese soldiers charge through the mist.

On February 23rd the platoon under Hank Hanson's command is ordered to climb Suribachi. They reach the top and hoist the American flag atop the mountain, and for the first time in a thousand years an enemy flag is raised on Japanese soil. Suddenly the platoon is attacked by Japanese sharpshooters, but the Marines kill them all.

Secretary of the Navy James Forrestal arrives on Iwo Jima and requests to have the flag atop Suribachi. Colonel Johnson is furious but ultimately gives in, telling Captain Severance to bring the flag down and replace it with another one. Severance sends Rene Gagnon, who is a runner, to go with Second Platoon to the top of the mountain and switch flags. When they reach the top they take down the first flag, and Mike, Harlon, Doc, Ira, Rene and Franklin raise the second. The event is seemingly insignificant, but it is captured by combat photographer Joe Rosenthal and the image goes on to live forever in the public consciousness.

On March 1st Second Platoon is on patrol when they are ambushed by a Japanese machinegun. Mike orders Harlon to have his parateam take out the machinegun nest, and the gunner is killed. Mike goes up to examine a dead Marine, and as he turns around and orders the unit to move out a Navy shell lands right behind him and knocks him down. In the smoke and confusion a Japanese soldier re-mans the machinegun and opens fire, killing the lieutenant and critically wounding Strank. Doc does everything he can, but Mike dies within minutes. Strank's death hits the squad hard, as they all idolized him, and things only get worse from then

Reel Marines

on. Later that day Hank is shot in the chest and dies almost instantly, and Harlon is killed by machinegun fire hours later. Two nights later, while Doc is helping a wounded Marine, Iggy is abducted by Japanese troops and dragged into a tunnel - and his mangled and tortured body is found a few days later by Doc. On March 21, as the battle is winding down, Franklin is killed by machinegun fire and dies in Ira's arms. Of the eight men in the squad only three are left - Doc, Ira and Rene. A few days after Franklin's death Doc is wounded by artillery fire while trying to save a fellow corpsman, but he survives and is sent back home. On March 26 the battle ends, and the Marines are victorious.

After the battle the press got hold of the photograph of the second flag raising. It is a huge morale booster, and papers all over the country ask for prints. When Rene is asked who is in the photo he gives five names, including his own, saying the other four are Mike, Doc, Franklin, and Hank, since he thought Hank was at the base of the flag when in reality it was Harlon. He then tells Ira he is the sixth man but Ira corrects him, saying that it was Harlon, and fiercely denies being in the photo - going so far as to threaten Rene with a bayonet to the throat. Even though Rene tells him they'll be sent home Ira won't give in, however when Rene is threatened with being sent back to the fighting he tells their bond tour guide that Ira was the sixth man.

Doc, who was in the hospital, is sent stateside with Ira and Rene as part of the seventh bond tour drive to raise money for the war effort. When they go to Washington they meet Bud Gerber of the Treasury Department, who will be their other guide. Doc notices that Hank's mother is on the list of mothers of the dead flag raisers, and Ira gets mad and calls the whole thing a farce. An annoyed Bud then confesses that the country is bankrupt and that if the bond drive fails the

Reel Marines

war will be lost. The three give in and decide not to tell anyone that Harlon Block was actually in the photograph.

The bond drive begins, and the three flag raisers are sent around the United States to raise money and make speeches. Ira gets drunk frequently, often breaking down from the memories that haunt him. On the night the three men raise a fake flag at Soldier Field Ira gets drunk, and throws up in front of General Alexander Vandergrift, Commandant of the Marine Corps. Vandergrift is furious and tells the guides to send Ira back to his unit. When Keyes tells Ira he's going back, Ira confesses that he can't stand being called a hero, and says Mike was the true hero. Ira says goodbye to Doc and Rene and goes back to the Pacific.

In September the war ends and Doc, Rene and Ira go home. Ira tries to move on but is never able to escape his unwanted fame, and one day in 1952 - after being released from jail - he hitchhikes over 1,300 miles to Texas to see Harlon Block's family. He tells Ed Block, Harlon's father, that Harlon was indeed at the base of the flag in the famous photograph. Then in 1954 the USMC War Memorial is dedicated, and the three flag raisers see each other one last time. In 1955 Ira dies of exposure after a night of drinking, and that same year Doc drove to the town where Iggy's mom lived and told her how Iggy died, although it is implied he lied. Rene has little success as the business offers he received on the bond drive are no longer offered to him, and he spends the rest of his life as a school janitor, dying in 1979. Doc is the only successful one. He buys the funeral home he worked at before the war and runs it for the rest of his life, and as he is on his death bed in 1994, he tells his son James how after the flag raising Captain Severance took the men swimming. He then dies peacefully, and in a final flashback to 1945 the men swim in the ocean after raising the flag.

Reel Marines

Trivia

❖ *Flags of Our Fathers* was directed, co-produced and scored by Clint Eastwood.

❖ The film *Letters From Iwo Jima* is a companion piece to *Flags of Our Fathers and* portrays the Battle of Iwo Jima from the perspective of the Japanese.

❖ *Flags of Our Fathers* is based on the book of the same name by James Bradley and Ron Powers about the Battle of Iwo Jima and the seven men who were involved in raising the flag on Mount Suribachi.

❖ A large part of the movie was filmed in Iceland.

❖ Actual Marines from the 5th Marine Regiment were used as extras during filming aboard ship as well as during the training work up, and can be best seen climbing up and down the cargo nets.

❖ The technical advisors on this film were retired Marine Sergeant Major Jim Dever and Master Sergeant Tom Minder, both of whom served in 1st Force Recon with the author of this book.

❖ The scene in which a sailor falls overboard and is left in the water as the fleet steams toward Iwo Jima actually happened. The incident is described in the book *Iwo* by Richard Wheeler, himself a veteran of the fighting. "According to Coast Guardsman Chet Hack of LST 763: 'We got the man-overboard signal from the ship ahead of us. We turned to port to avoid hitting him and threw him a life preserver, but had orders not to stop. We could not hold up twenty-four ships for one man. Looking back, we

Reel Marines

could see him waving his arms, and it broke our hearts that we couldn't help him. We hoped that one of our destroyers or other small men-of-war that were cruising around to protect us would pick him up, but we never heard that they did.'"

❖ At the 2008 Cannes Film Festival director Spike Lee, who was ironically making the film *Miracle at St. Anna* about an all-black U.S. division fighting in Italy during World War II, criticized Clint Eastwood for not depicting black Marines in *Flags of Our Fathers*. Citing historical accuracy, Eastwood responded that his film was specifically about the Marines who raised the flag on Mount Suribachi at Iwo Jima, pointing out that while black soldiers did fight at Iwo Jima, the U.S. military was segregated during WWII and none of the men who raised the flag were black. Eastwood angrily said that Lee should "shut his face." Lee responded that Eastwood was acting like an "angry old man," and argued that despite making two Iwo Jima films back to back, *Letters from Iwo Jima* and *Flags of Our Fathers*, "there was not one black Marine in either film."

❖ In actual fact black Marines (including an all-black unit) are seen in scenes during which the mission is outlined, as well as during the initial landings when a wounded black Marine is carried away, and during the end credits historical photographs taken during the battle show them as well. Although black Marines fought in the battle, they were restricted to auxiliary roles such as ammunition supply and were not involved in major assaults.

Reel Marines

FLYING LEATHERNECKS

Release date: 28 August 1951
Running time: 102 minutes
Historical context: World War II

Tagline: From Guadalcanal to Okinawa... the Marine air-devils blazed a trail of glory.

Cast

John Wayne - Major Daniel Xavier Kirby
Robert Ryan - Captain Carl "Griff" Griffin
Don Taylor - Lieutenant Vern "Cowboy" Blithe
Jay C. Flippen - Master Sergeant Clancy

Quote: "For the future record, all orders are right away."
- Major Dan Kirby

Highlights

What more can you ask for... John Wayne shaping up a bunch of cocky pilots, real-life Marine Robert Ryan as his XO, and a lot of aerial combat action. Marines are taught the birth of Close Air Support took place at Bougainville, but won't mind the literary license which shows it happening on Guadalcanal. Add to the mix Jay C. Flippen as Master Sergeant Clancy, who could be the model for every unit "scrounger," and you have the formula for a thoroughly entertaining film.

Reel Marines

Plot

Major Dan Kirby arrives at VMF-247 as the new commander while everybody in the unit was expecting Captain Carl "Grif" Griffin to get the job. Kirby is strict about rules being followed, and makes it understood from day one. While assigned to the "Cactus Air Force" during the Guadalcanal campaign, Kirby has few planes available and a lot to accomplish on an airfield which is attacked daily by the Japanese. His pilots are young and behave like "kids," at times disobeying orders and foolishly losing their lives and precious planes. Kirby is requiring maximum effort, and Captain Griffin is not as tough as Kirby expects. Griffin stays closer to his young pilots - one of them being his own brother-in-law, Vern "Cowboy" Blithe.

Kirby for his part hates the decisions he has to make and knows he is sending his pilots to their deaths, but the success of the mission is the most important thing to him. He keeps this a secret from the rest of his squadron, as the hard conditions of the war force Kirby to get all the more strict with his exhausted pilots. He even refuses sick leave to men with malaria, and does not allow them to return to base when planes have technical problems. The tension between Griffin and Kirby soon comes to a head, and they have a hard discussion. It is obvious Griffin does not recognize the hardships Kirby faces, because he is often more driven by his sentimental side.

Kirby is a fan of low-level ground attacks to support Marine infantry units, but HQ does not approve of his tactics until the time comes when Marines are pinned down by the Japanese. Kirby then adjusts his squadron's tactics accordingly and, despite losing a number of pilots, tries to prove his point. In his most successful operation he leads his squadron in an attack on a huge Japanese convoy - a scenario most likely based on the Naval Battle of Guadalcanal. Now promoted to Lieutenant Colonel, Kirby is given the chance to develop low level attack tactics in the States.

Kirby returns to the front and to the same unit and crew, now equipped with F4U Corsair fighters. He leads his men against Japanese troops and Kamikaze attacks during the Battle of Okinawa, and during a crucial moment Griffin denies assistance to his brother-in-law to avoid splitting his formation, and as a result Cowboy is killed. Kirby shortly thereafter is shot down and injured, but is picked up by a Navy launch. Since he is now to leave the squadron he has to appoint a successor, and he appoints Griff as CO because he believes Griffin is now able to put the lives of his pilots second to his duty and they split with a friendly promise to meet again. Kirby admits that every moment in which he is required to make a decision is a nightmare, but that comes with the territory of being a leader.

Throughout the film the character of Sergeant Clancy (Jay C. Flippen), an old Marine veteran and friend in arms for Kirby, spices up scenes with some fun. Clancy is creative in getting provisions for his outfit, much to the consternation of other units on the island due to his unorthodox methods. His "improvising" may furnish the ideal solutions for the poorly equipped VMF-247, but by the end of the film Clancy will be caught by the MPs and lose some stripes.

Trivia

❖ This movie is often considered another assignment of Nicholas Ray's at RKO for Howard Hughes, to prove his political and professional allegiance during the Red Scare.

Reel Marines

- Robert Ryan enlisted in the Marine Corps during World War II and served as a drill instructor at Camp Pendleton.

- Director Nicholas Ray chose Ryan to play opposite John Wayne because he had been a champion boxer in college and was the only actor Ray could think of who could "kick Wayne's ass."

- The fighter planes appearing in the first part of the film are not the historically accurate Grumman F4F Wildcats but instead F6F Hellcats. Apparently few Wildcats had survived, while an appreciable number of Hellcats were available in 1951 when the film was produced.

- Hellcats painted white were used as Japanese Zeros.

- The role of Major Kirby was inspired by real World War II flying ace Major John L. Smith for his missions over Guadalcanal in 1942. His actions in the war were renowned by the time the film was made and he was awarded the Medal of Honor in 1943 and later promoted to Lieutenant Colonel, as was Kirby in the film. There is a distinct similarity in appearance between Smith and John Wayne.

- The squadron Smith had actually flown with was VMF-223, while the one in the film carried the designation VMF-247. VMA 223 has been active since 1st May 1942, and is currently comprised mainly of AV-8B Harrier jump jets.

- Ryan and Ray, who were leftist liberals, constantly fought against John Wayne and Jay C. Flippen, both of whom were conservatives and supported the Blacklist.

Reel Marines

FULL METAL JACKET

Release date: 26 June 1987
Running time: 116 minutes
Historical context: Vietnam

Tagline: In Vietnam the wind doesn't blow, it sucks.

Cast

R. Lee Ermey - Gunnery Sergeant Hartman
Matthew Modine - Private Joker / Narrator
Vincent D'Onofrio - Private Pyle
Adam Baldwin - Animal Mother
Arliss Howard - Private Cowboy
Dorian Harewood - Eightball

Quote: "You talk the talk. Do you walk the walk?" - Animal Mother

Highlights

Full Metal Jacket is a "must see" for a number of reasons. First and foremost the boot camp scenes are nothing short of amazing – and because viewers are not subject to punishment, it's okay to laugh at R. Lee Ermey's lines. If this isn't the most quoted movie of all time, it surely is close. We all have known a Private Pyle, a Private Joker and a Gunny Hartman, and this film allows us to reflect on the unique personalities and crazy circumstances which make the Marine Corps what it is.

Reel Marines

Plot

During the Vietnam War a group of Marine Corps recruits arrive at Parris Island for boot camp. After having their heads shaved they meet their drill instructor, Gunnery Sergeant Hartman, who verbally and physically abuses the young men with the intention of desensitizing and hardening them. With the Vietnam War in full swing he has the task of producing trained warriors, and Hartman targets much of his abuse towards Privates Joker and Cowboy, and is especially tough on Private Pyle because he is overweight and mentally slow.

Because he is unresponsive to Hartman's continual negative reinforcement Pyle is paired up with Joker, who attempts to square him away. Thanks to Joker's patience and encouragement Pyle begins to improve, but things derail when Hartman discovers a contraband jelly doughnut in Pyle's footlocker which causes him to punish the entire platoon for Pyle's failings. As a result the platoon hazes Pyle one night by pinning him to his bunk with a blanket and beating him with bars of soap wrapped in towels. Joker reluctantly joins in, but then out of frustration beats Pyle several times. The next day Pyle is transformed by this event, becoming a model Marine and expert rifleman, but shows signs of mental breakdown including social withdrawal and talking to his M14 rifle.

After graduating each recruit is assigned a Military Occupational Specialty, with most being sent to the Infantry, although Joker is assigned to Basic Military Journalism. On the platoon's last night on Parris Island Joker draws fire watch, during which he discovers Pyle in the head loading his rifle with live ammunition. Frightened, Joker attempts to calm him, but Pyle begins shouting, executing drill

Reel Marines

commands, and reciting the Rifleman's Creed. The noise awakens Hartman, who confronts Pyle and demands that he surrender. Instead Pyle turns the rifle on Hartman, murders him, and then places the muzzle in his own mouth and commits suicide.

Sometime later in January of 1968 Joker has become a Corporal and a Marine Combat Correspondent in Vietnam with *Stars and Stripes*, and is assigned to a Marine public-affairs unit along with PFC Rafterman (Kevyn Major Howard), a combat photographer. Rafterman wants to go into combat, as Joker claims he has been, although one of his colleagues mocks him and states he knows Joker has never been in combat because he doesn't have the "thousand-yard stare." Their argument is interrupted by the sound of nearby gunfire, as the North Vietnamese Army has begun the Tet Offensive and is attempting to overrun the base.

The next day the staff learns about enemy attacks throughout South Vietnam. Joker's commander assigns him to Phu Bai, a Marine forward operating-base near the ancient Vietnamese city of Huê, to cover the combat taking place there, and Rafterman tags along hoping to get some combat experience. When they land outside Huê they meet up with the Lusthog Squad, where Cowboy is now a Sergeant and second-in-command. Joker accompanies the squad during the Battle of Huê, during which their commander, Lieutenant Touchdown (Ed O'Ross) is killed and another Marine nicknamed Crazy Earl takes command.

A few days later the squad goes out on patrol again, this time north of the Perfume River which divides the city of Huê, where the Americans believe enemy forces have hidden. Crazy Earl comes across a toy rabbit in a ruined building and picks it up, triggering an explosive booby trap which kills him and leaves Cowboy as the reluctant squad leader. The squad becomes lost in the ruined buildings and a sniper pins them down, while wounding two of their comrades. The sniper refrains from killing the wounded men, with the apparent intention of drawing more Marines into the killing zone. As the squad maneuvers to try to locate the hidden position, Cowboy is shot and killed as well.

With Cowboy dead, an M-60 Machine gunner named Animal Mother assumes command of the remaining Marines. Using smoke grenades to conceal their advance, the squad locates the sniper. Joker finds the sniper on an upper floor, but his rifle jams as he tries to fire. The sniper, a young girl, spins around, opens fire and pins him behind a column. Joker frantically drops his rifle and draws his sidearm, but Rafterman arrives and shoots the sniper, saving him. As Animal Mother and other Marines of the squad converge she begins to pray and then repeats "shoot me," prompting an argument about whether to leave her to die from her wounds or put her out of her misery. Animal Mother decides to allow a mercy killing only if Joker performs it, and after some hesitation Joker shoots her with his sidearm. The Marines congratulate him on his kill as Joker stares into the distance, and the film concludes with the Marines marching toward their bivouac singing the Mickey Mouse March. Joker states that, despite being "in a world of shit," he is glad to be alive, and is unafraid.

Trivia

❖ *Full Metal Jacket* is based on the novel *The Short-Timers* by former Marine Combat Correspondent Gustav Hasford. Kubrick discovered it while reading the *Virginia Kirkus Review*.

Reel Marines

- The title refers to the full metal jacket ammunition used by infantry riflemen, as heard in Private Pyle's famous line spoken in the head, "7.62 millimeter, full... metal... jacket." Full metal jacket ammunition has a copper coating covering the lead core of its projectile.

- Stanley Kubrick contacted Michael Herr, author of the Vietnam War memoir *Dispatches*, in the spring of 1980 to discuss working on a film about the Holocaust, but eventually discarded that idea in favor of a film about Vietnam.

- Herr was not initially interested in revisiting his Vietnam War experiences, and Kubrick spent three years persuading him in what the author describes as "a single phone call lasting three years, with interruptions."

- According to the stenciling on his sweatshirt, Private Joker's real name is J.T. Davis.

- The humiliating nickname Gomer Pyle refers to a likable but dim-witted character from *The Andy Griffith Show* who eventually enlists in the USMC.

- Lieutenant Walter J. "Touchdown" Schinowski got his nickname because he was a college football player at the University of Notre Dame.

- Kubrick was worried that the title of the book *Short-Timers* would be misread by audiences as referring to people who only did half a day's work and changed it to *Full Metal Jacket* after discovering the phrase while going through a gun catalog.

- At some point Kubrick wanted to meet Hasford in person but Herr advised against it, describing *The Short-Timers*

69

Reel Marines

author as a "scary man." Kubrick insisted, and they all met at Kubrick's house in England for dinner. It did not go well, and Hasford was subsequently shut out of the production.

❖ Former Marine Drill Instructor R. Lee Ermey was originally hired as a technical adviser and asked Kubrick if he could audition for the role of Hartman, but Kubrick, having seen his portrayal of Drill Instructor SSgt Loyce in *The Boys in Company C*, told him he wasn't vicious enough to play the character. In response Ermey made a videotape of himself improvising insulting dialogue towards a group of Royal Marines while people off-camera pelted him with oranges and tennis balls. Despite the distractions Ermey rattled off an unbroken string of insults for fifteen minutes and did not flinch, duck, or repeat himself while the projectiles rained on him. Upon viewing the video Kubrick gave Ermey the role, realizing "he was a genius for this part."

❖ Ermey's experience as a real-life DI during the Vietnam era proved invaluable, and he fostered such realism that in one instance he barked an order off-camera for Kubrick to stand up when he was spoken to and Kubrick instinctively obeyed - and was standing at attention before realizing what had happened.

❖ Kubrick estimated that Ermey came up with 150 pages of insults, many of them improvised on the spot - a rarity for a Kubrick film. Overall the former drill instructor wrote 50% of his own dialogue, especially the insults.

❖ Ermey usually needed only two to three takes per scene, another rarity for a Kubrick film.

Reel Marines

❖ The original plan envisaged Anthony Michael Hall starring as Private Joker, but after eight months of negotiations a deal between Stanley Kubrick and Hall fell through.

❖ Kubrick shot the film in England, in Cambridgeshire on the Norfolk Broads, and at the former Beckton Gas Works in Newham (East London), with the disused Gasworks doubling as the ruined city of Huế. Kubrick worked from still photographs of Huế taken in 1968, and had buildings blown up and used a wrecking ball to knock specific holes in others over the course of two months.

❖ Bassingbourn Barracks, a former RAF and British Army base, doubled as the Parris Island Marine boot camp. A British Army rifle range near Barton was used in the scene where Private Pyle is congratulated on his shooting skills.

❖ Kubrick acquired four M41 tanks from a Belgian army colonel, Sikorsky H-34 Choctaw helicopters (actually Westland Wessex painted Marine green), and obtained a selection of rifles, M79 grenade launchers and M60 machine guns from a licensed weapons-dealer.

❖ Matthew Modine described the shoot as tough, since he had to have his head shaved once a week and Ermey yelled at him for ten hours a day during the shooting of the Parris Island scenes.

❖ At one point during filming Ermey had a car accident, broke all of his ribs on one side, and was out for four-and-half months. As a result of his injuries, in some scenes he does not move his left arm at all.

Reel Marines

- Hartman is never seen without his cover, including the scene where he is wearing skivvies in the head.

- According to director John Boorman, Stanley Kubrick wanted to cast Bill McKinney in the role of Hartman but was so unsettled after viewing McKinney's performance as "Mountain Man" in *Deliverance* that he declined to meet with him, saying he was simply too frightened at the idea of being in McKinney's presence.

- Kubrick hired Tim Colceri to play Hartman after discarding his original idea of casting Bill McKinney, but he never got to play the role because Kubrick decided to use Ermey instead. Colceri was bitter, but accepted Kubrick's consolation prize of a small role as the helicopter door-gunner.

- Vincent D'Onofrio gained seventy pounds for his role as Pyle, breaking Robert De Niro's movie weight-gain record of sixty for Raging Bull in 1980. It took him seven months to put the weight on, and nine to take it off.

- Mickey Mouse is referred to at the end of both segments: when Hartmann enters the head to confront Joker and Pyle he screams, "What is this Mickey Mouse shit?" and Joker and company later sing the theme from the *Mickey Mouse Club* as they march through the burning city. A third Mickey Mouse reference is in the press room, where a Mickey Mouse figure can be seen near the window behind Joker.

- While filming the opening scene where Hartman disciplines Private Cowboy, he says Cowboy is the type of guy who would have sex with another guy "and not even have the goddamned common courtesy to give him

Reel Marines

a reach-around." Kubrick immediately yelled cut and asked Ermey, "What the hell is a reach-around?" Ermey explained, and Kubrick laughed and re-shot the scene - telling Ermey to keep the line.

❖ The inscription "I Am Become Death" written on Animal Mother's helmet is a quotation from the *Bhagavad-Gita*, and was spoken by J. Robert Oppenheimer after the explosion of the first atomic bomb at Alamogordo.

❖ The only shot that actually shows Parris Island is when the platoon graduates and a piece of video imported into the movie shows a graduation taking place in the First Battalion area.

❖ Advertisements for this film were censored in some parts of Canada due to the tagline "In Vietnam the wind doesn't blow, it sucks." At that time Canadian censors had not yet decided whether the phrase was obscene.

❖ To make Hartmann's performance and the recruits' reactions as convincing as possible, the actors playing recruits never met R. Lee Ermey prior to filming. Stanley Kubrick also saw to it that Ermey didn't fraternize with them between takes.

❖ One scene cut from the movie showed a group of soldiers playing soccer. It was cut because they were not using a soccer ball, but a human head.

❖ Kubrick shot a scene in the Norfolk Broads where a "Wessex" helicopter flown by a stunt pilot was required to fly low along a canal doubling as paddy fields while someone fired a machinegun out of the doors. The scene was shot at dawn, and the local police were supposed to

Reel Marines

have warned fishermen - but there was a communications problem, and many fishermen thought there was a U.S. helicopter machine-gunning their boats.

❖ Dorian Harewood visited a doctor twice while shooting his scenes, fearing his eardrum had been blown out by Ermey's screaming.

❖ The exchange between the Da Nang Hooker and Joker, "me so horny, me love you long time," was sampled in *2 Live Crew's* 1989 hit song *Me So Horny* on the album *As Nasty As They Wanna Be.*

❖ Ermey actually slapped Vincent D'Onofrio in the scene where he knocks his cover off. It was D'Onofrio's idea, but unfortunately for him he forgot about Kubrick's perfectionism and had to endure take after take of real slaps.

❖ Val Kilmer auditioned for the part of Joker. According to Matthew Modine, Kilmer confronted him in a restaurant and challenged Modine to a fight because he believed he had stolen the part from him - but Modine was not even aware of the film at the time. Modine later sent Kubrick footage of his performance in *Vision Quest* and got the part.

Reel Marines

GENERATION KILL

Release date: 13 July 2008
Running time: 63–71 minutes (seven episodes)
Historical context: Operation Iraqi Freedom

Tagline: Stay frosty!

<u>Cast</u>

Alexander Skarsgård - Sergeant Brad "Iceman" Colbert
James Ransone - Corporal Josh Ray Person
Lee Tergesen - Evan "Scribe" Wright
Jon Huertas - Sergeant Antonio Espera
Stark Sands - First Lieutenant Nathaniel Fick
Billy Lush - Lance Corporal Harold James Trombley
Jonah Lotan - HM2 Robert Timothy "Doc" Bryan
Rudy Reyes - Sergeant Rudy Reyes

Quote: "Hey Walt, can you keep it down? I'm having trouble hearing the artillery." - Corporal Person

Highlights

Generation Kill is both real and contemporary, and it is a truly important film because it fleshes out and personalizes the too-often vague media reports about the war in Iraq. In order to truly understand this conflict and the nature of the enemy it is necessary to view things through the lens of the grunts who are fighting on the ground, in addition to the macro-strategic overview we get from the press.

Reel Marines

Plot

The mini-series premiered on July 13, 2008 and spanned seven episodes. The First Reconnaissance Battalion was commanded by Lieutenant Colonel Stephen "Godfather" Ferrando, played by Chance Kelly, while Bravo Company was under the command of Captain Craig Schwetje (played by Brian Patrick Wade), and Bravo Company's third platoon was commanded by the erratic Captain Dave "Captain America" McGraw (played by Eric Nenninger). The following synopsis breaks down the series episode by episode.

In the episode "Get Some" the Marines prepare to invade Iraq at the beginning of Operation Iraqi Freedom. While waiting to receive their orders at Camp Mathilda in Kuwait they learn that *Rolling Stone* columnist Evan Wright will be embedded with them.

In "The Cradle of Civilization" the invasion of Iraq is in full swing, and Sergeant Colbert tries to keep his unit focused. First Recon Marines adjust to shifting attack plans while anticipating their first contact with the enemy in Nasiriyah and Al Gharraf.

In "Screwby" Bravo Company awaits their orders for a recon mission after having survived its first trial by fire, Encino Man requests an artillery strike on a phantom RPG team, and Fick tries to take control of a dangerous situation. Lieutenant Colonel Ferrando then issues a new, more urgent order shortly after Alpha Company shells Ar Rifa.

"Combat Jack" depicts grumbling in the ranks about an abandoned supply truck at the captured airfield, but Bravo is soon on the move again and heading north, clearing villages and setting up a roadblock outside Al Hayy. Meanwhile,

Reel Marines

Alpha is ordered to find the body of a Marine in Al Shatra, but their mission is delayed by a CIA operation.

"A Burning Dog" reveals that despite an armored division's punishing response to First Recon's intelligence-gathering about an ambush-in-waiting at a strategic bridge, Bravo still meets stiff resistance while making several attempts to cross it. A survey of the battlefield prompts more questions than answers about the enemy, and a roadblock in Al Muwaffiqiyah tests the Marines' ever-changing rules of engagement.

In "Stay Frosty" First Recon is assigned the unfamiliar mission of escorting hundreds of civilians fleeing Baghdad, and they begin to wonder if their part in the war may be ending. Lieutenant Colonel Ferrando has other plans to get his men back into the battle.

In the final episode, "Bomb in the Garden," Bravo Company reaches Baghdad and is shocked at the size of the city. When First Recon begins doing their daily patrols in Baghdad, they find out the obstacles they and the Iraqis face are much greater than they could have ever imagined.

Trivia

❖ *Generation Kill* is a 2008 HBO television mini-series based on the book of the same name by *Rolling Stone* reporter Evan Wright. It is based on his experiences as an embedded reporter with the 1st Reconnaissance Battalion during the Iraq War's first phase in 2003.

❖ The majority of the characters were drawn from the Second Platoon of the First Reconnaissance Battalion's Bravo Company.

Reel Marines

- Wright was assigned to the lead vehicle of Bravo Company, which he shared with Sergeant Colbert, Corporal Person, and Lance Corporal Trombley.

- Throughout the entire series Wright is never referred to by his name. He is instead called "Scribe," "Rolling Stone," or "Reporter."

- Rudy Reyes, the muscular Marine who runs around the camp with a gas mask on, portrays himself.

- Marine Eric Kocher served as the series Key Military Advisor. He does not portray himself in the series, but instead appears as Gunnery Sergeant Rich Barrett.

- The unusual tattoo on Sergeant Colbert's back appears to be based on a landscape done by Spanish artist Luis Royo.

- The film was shot on location in South Africa, Namibia and Mozambique.

Reel Marines

Release date: 27 October 1943
Running time: 93 minutes
Historical context: World War II

Tagline: A war correspondent telling the story of how U.S. Marines fought and died at Guadalcanal.

Cast

Preston Foster - Father Donnelly
Lloyd Nolan - Gunnery Sergeant Hook Malone
William Bendix - Corporal Aloysius T. "Taxi" Potts
Richard Conte - Captain Don Davis
Anthony Quinn - Private Jesus "Soose" Alvarez
Richard Jaeckel - Private Johnny "Chicken" Anderson

Quote: "As soon as our planes get here they'll be sorry."
- Corporal Potts

Highlights

The three things which stand out about Guadalcanal Diary are the personal stories of the Marines, the historical context in which the story unfolds, and the authenticity of the combat scenes. Another high point is the appearance of a real hero of this battle in a cameo role. This film was made at a dark time in America's history, and played a very real role in rallying the Nation!

Reel Marines

Plot

Guadalcanal Diary tells the story of the United States Marines during the Battle of Guadalcanal, which occurred only a year before the movie's release and would ultimately turn the tide of the war. While the film has notable battle scenes, its primary focus is on the characters and back stories of the Marines. The film begins on a beautiful sunny day while the men rest and wait and wonder where they are going. At that time you are introduced to the characters, and learn just who these men really are. The tense moments before landing show the men preparing for the unknown, each in their own way. Once they land on the beachhead, the film shows the Marines' journey as they try to take and hold the island.

When the anxious and unsuspecting Marines first land in the Solomon Islands they must adapt and quickly learn how to engage the Japanese in jungle warfare. The story follows one squad of Marines through the bloody assaults on the Solomon Islands during the opening stages of the war in the South Pacific. There's the tough sergeant (Lloyd Nolan), a cab driver from Brooklyn (William Bendix), a Mexican (Anthony Quinn) and a chaplain (Preston Foster).

Throughout the movie a battle-weary narrator reads from a diary, commenting on the typical grunt's everyday life - and when necessary, his death. Battles and dates of engagement are named to put the explosive action into a solid historical context, and those details provide the authenticity and historical perspective which make *Guadalcanal Diary* more than just another war movie.

Reel Marines

Trivia

- *Guadalcanal Diary* is based on the book of the same name by Richard Tregaskis.

- Marine Corps Captain Marion Carl, a multi-ace with 18.5 aerial victories, makes an appearance as a Marine pilot. He wears his baseball cap with the bill pointed skyward and makes the comment, "Don't look now fellas, but a truck of gas just came on the field." Carl was a survivor of the Battle of Midway and the air campaign for Guadalcanal in 1942, and was awarded two Navy Crosses. Sadly, on June 28, 1998, he was murdered in his Oregon home by an intruder.

- The movie accurately shows the Marines armed with Springfield bolt action rifles. Garand rifles were available, just not in the numbers necessary to issue before setting sail.

- This was the movie debut of Richard Jaeckel, who was just seventeen years of age at the time. He was a messenger for 20th Century-Fox studios when he was cast in the film.

- This movie was made in 1943, only one year after the initial landing on Guadacanal, and premiered about ten months after the end of the campaign.

- Guadacanal was named after Pedro de Ortega's hometown of Guadacanal in Andalusia, Spain. de Ortega worked under Álvaro de Mendaña, who charted the island in 1568.

Reel Marines

- ❖ Captain Clarence Martin, who fought with the first detachment of Marines at Guadalcanal, acted as a technical adviser on the movie.

- ❖ The Hollywood premiere of *Guadalcanal Diary* was a charity benefit to aid various war charities with the sixty-piece Pendleton Field Marine Band performing at the bash. According to the *Hollywood Reporter*, the launch was attended by "top-ranking officers of the Marines, Army and Navy... (and) about fifty war heroes."

- ❖ The Philadelphia Premiere for this movie was dedicated to celebrating the 168th Anniversary of the inception of the United States Marine Corps.

- ❖ William Bendix told the *Saturday Evening Post's* "The Role I Liked Best" column in 1946 that his character of Corporal Aloysius T. "Taxi" Potts was his favorite as it had given him "the widest range of opportunity" for an actor. Moreover, Bendix stated that he was moved by the letters he received from military personnel who recognized his gutsy performance as a Marine in this movie. Bendix added that he and his fellow cast members enjoyed the experience of working with the Marines based at Camp Pendleton.

- ❖ Crew member Donald Petersen suffered broken ear drums from dynamite which exploded prematurely, and was awarded $15,000 in damages in a jury trial which was then appealed by the studio.

Reel Marines

GUNG HO!

Release date: 20 December 1943
Running time: 88 minutes
Historical context: World War II

Tagline: Battle cry of the Marine Raiders!

Cast

Randolph Scott - Colonel Thorwald
Alan Curtis - John Harbison
Noah Beery Jr. - Corporal Kurt Richter
J. Carrol Naish - Lieutenant C.J. Cristoforos
Richard Lane - Captain Dunphy
Walter Sande - "Gunner" McBride
Robert Mitchum - "Pig-Iron" Matthews

Quote: "A guy could get killed in here!" - Corporal Richter

Highlights

The Makin Island Raid is one of the most famous actions in Marine Corps history, and its on-screen depiction serves to ensure this story will be remembered by generations to come. While the movie version does of course take some creative license with the actual events surrounding the raid – and many of the characters were fictionalized because the war was still going on at the time of filming – the script and quasi-biographical portrayals do an excellent job of capturing the spirit of the Marine Raiders.

Reel Marines

Plot

The film begins with a tough Greek Lieutenant (J. Carroll Naish) announcing the Marine Corps is seeking volunteers for a hazardous mission and special unit. Sergeant "Transport" Anderof (Sam Levene) meets the commander of the unit, Lieutenant Colonel Thorwald (Randolph Scott) with whom he has served in the China Marines. Thorwald explains that he left the Corps to serve with the Chinese guerillas fighting the Japanese during the Second Sino-Japanese War to learn their methods, and has decided to form a unit using what he learned.

Amongst the volunteers for the unit are a hillbilly (Rod Cameron) - who responds to the Marine Gunner's (Walter Sande) question about whether he can kill someone by revealing that he already has - and ordained Minister Alan Curtis, who is keeping his vocation a secret. Robert Mitchum is "Pig Iron," a boxer from a background of poverty and hard work. Harold Landon is a young, small street kid who is initially rejected by Naish, but wins him over because both worked as dishwashers on ships bound to the United States from Pireaus. Volunteers with brief screen time include a Filipino wishing to avenge his sister who was raped and killed in Manila and teaches the Raiders knife fighting, a veteran of the Spanish Civil War who sees the war as a continuation of the fight against Fascism, and a Marine who honestly admits, "I just don't like Japs."

The film moves rapidly in a documentary style with stock footage of training narrated by Chet Huntley. The survivors of the training are sent to Hawaii for further jungle warfare exercises, where they witness the damage of the attack on Pearl Harbor. In Hawaii they hear a radio bulletin of the announcement of the Battle of Guadalcanal. The Marines are then ordered to board two submarines, *USS Nautilus* and *USS Argonaut,* destined for a commando raid on a Japanese held island.

After a claustrophobic voyage, the Raiders infiltrate the island in rubber boats. The landing is met by fire from snipers hiding in palm trees, but the Marines dispose of them, attack the Japanese headquarters, wipe out the Japanese garrison, destroy installations with explosives, and then board their submarines to return home.

Thorwald lectures throughout the film that the Japanese have no initiative and cannot think for themselves or deviate from a plan, thus unexpected action pays off. This is demonstrated in several scenes where a Marine defeats his opponent in unarmed combat by spitting tobacco in his opponent's eyes, a small but fast runner strips down to his trousers and quickly zigzags through enemy fire to deliver hand grenades, Marines destroy a Japanese pillbox and its occupants by squashing both with a road construction steamroller, and when a speechless Robert Mitchum - who has been shot in the throat and is unable to give warning - kills a Japanese infiltrator attempting to kill the battalion surgeon (Milburn Stone) by throwing his knife into the enemy soldier's back. The climax of the film has the Raiders painting a giant American flag on the roof of a building, and then luring the counterattacking Japanese to the area - where their own air force bombs and strafes them.

In contrast to the Japanese and the rest of the American military, Thorwald orders his officers to wear no rank insignia and have no special privileges, and tells his Raiders "I will eat what you eat, sleep where you sleep, and participate in the same training." His Raiders then participate in "Gung Ho Sessions" where they discuss the unit's plans, and each man participates without regard to rank.

Reel Marines

Trivia

❖ The story is based on the real-life World War II Makin Island raid led by Lieutenant Colonel Evans Carlson's 2nd Marine Raider Battalion.

❖ The phrase "Gung Ho," which literally means "working together," entered the public lexicon both from the film and the accounts of actual Raiders.

❖ When producer Walter Wanger acquired the rights of the Makin Island raid and Lieutenant W.S LeFrancois' story, Navy film liaison Lieutenant Albert Bolton insisted that neither Carlson nor his executive officer James Roosevelt be singled out - so the screenplay depicted a fictional "Colonel Thorwald" with no executive officer.

❖ The screenplay did include a character of Greek extraction played by J. Carroll Naish based on Marine Raider Lieutenant John Apergis, as well as Gunnery Sergeant Victor "Transport" Maghikian, who served in the raid and survived the war.

❖ Though many incidents depicted in the film did not occur in the real Makin Island raid, Carlson wrote to Wanger that he was pleased with the film.

❖ Like many films about the Corps, this movie was filmed at Marine Corps Recruit Depot San Diego and Camp Pendleton with Marine extras and technical advisors.

❖ Since we were at war, the Japanese were necessarily played by Chinese and Filipino extras.

Reel Marines

❖ Many individuals who accuse the Marine Corps of deliberately recruiting murderers and criminals may have been inspired by Rod Cameron's role in the film.

❖ In the early 1960s Louis Marx and Company came out with a "Gung Ho Commando Outfit" for children.

❖ Shortly after the United States entered World War II Randolph Scott attempted to obtain an officer's commission in the Marines Corps, but due to a back injury from years earlier he was turned down.

❖ Scott had previously appeared as a Marine in the 1941 movie *To the Shores of Tripoli.*

Reel Marines

HALLS OF MONTEZUMA

Release date: 4 January 1951
Running time: 113 minutes
Historical context: World War II

Tagline: Action with the Battling Leathernecks!

<u>Cast</u>

Richard Widmark - Lieutenant Carl Anderson
Jack Palance - Pigeon Lane
Reginald Gardiner - Sergeant Randolph Johnson
Robert Wagner - Private Coffman.
Karl Malden - PHM2 C.E. "Doc" Jones
Richard Boone - Lieutenant Colonel Gilfillan
Jack Webb - Correspondent Sergeant Dickerman
Neville Brand - Sergeant Zelenko
Martin Milner - Private Whitney

Quote: "General Sherman said, 'War is Hell.' What did he know, that eight-ball never left the States." - Lt. Butterfield

Highlights

Halls of Montezuma has a great cast, and many of its stars would go on to have great success in Hollywood. The plot, which emphasizes the necessity for gathering intelligence, adds interest to the standard frontal assaults which fill war films. Richard Widmark's character is key, and his burden is one with which is carried by many "citizen-soldiers."

Reel Marines

Plot

Halls of Montezuma is a 1951 World War II action/drama which follows a group of U.S. Marines from the beach to a Japanese rocket site through enemy infested jungles as their ex-school teacher leader is transformed into a battle veteran and his squad becomes a tight fighting unit.

During World War II, as a Marine battalion prepares to land on a large Japanese-held island in the South Pacific, Lieutenant Colonel Gilfillan warns his men that it will be a tough mission and that they have been ordered to take prisoners in order to gain information about the Japanese fortifications. Below decks Lieutenant Carl Anderson, a chemistry teacher in civilian life, questions his former student, Corporal Stuart Conroy, who complains he is ill and cannot fight. Anderson assures him he has shown courage before, and can do so again. In the landing boat heading to shore, Corpsman "Doc" Jones is worried because Anderson has been suffering from "psychological migraines" for months. Although Doc urged Anderson to seek treatment in the United States, he refused to leave his men and has been relying on Doc to supply him with painkillers.

The men hit the beach and successfully dig in, despite an initial burst of resistance. As four days pass Anderson's squad, which includes boxer Pigeon Lane, Sergeant Zelenko, Slattery, Coffman, Whitney and the unstable "Pretty Boy" Riley, grows weary of the constant threat of hidden Japanese snipers. One day the men try to take a ridge, but are beaten back by Japanese rockets which come as an unpleasant surprise to the commanding officer. When Coffman is killed, Anderson is forced to take some more of Doc's pills.

Anderson later meets with other officers at headquarters, where Gilfillan recounts the troubles they are having

Reel Marines

capturing prisoners and getting information from them. Sergeant Randolph Johnson, an unconventional Marine who specializes in tricking Japanese prisoners into talking, questions a prisoner who has been dubbed "Willie." As Gilfillan receives orders to stop the rockets within nine hours, before the next assault on the hills, Willie informs Johnson that Japanese soldiers holding a cave stronghold are willing to surrender. Accompanied by Johnson and war correspondent Sergeant Dickerman, Anderson leads a patrol to the cave but they are ambushed and Zelenko is blinded.

    The men capture the remaining Japanese, including Captain Ishio, Makino, Romeo, Nomura, and a shell-shocked, elderly civilian. Anderson finds a map on one of the dead officers, and then leads his men across a river and through the jungle. After a sniper shoots at him Pretty Boy kills the man during hand-to-hand combat, but the confrontation further unbalances him and he attempts to murder the prisoners. Lane then accidentally shoots and kills Pretty Boy while attempting to stop him. Doc also dies, but not before giving Dickerman a message for Anderson.

    Anderson and the remaining men return to headquarters, where Makino commits hara-kiri with a knife he had stolen from Johnson. While map expert Lieutenant Butterfield works on a Japanese overlay found by Pretty Boy, Anderson and Johnson learn Nomura is actually an important officer named Major Kenji Matsuoda. Johnson finally deduces where the rockets are located, and a barrage begins as Anderson rejoins his men outside. Anderson learns Conroy has been killed by a sniper, and with only Lane, Whitney and Slattery left of his original squad he takes the news hard and is ready to give up. Dickerman then reads aloud Doc's note and Anderson is inspired, causing him to throw away his painkillers and lead his men into battle.

Reel Marines

Trivia

- The movie's title *Halls of Montezuma* represents the first line of the *Marines' Hymn*. This song is the oldest official song in the United States military. The second line, *To the Shores of Tripoli,* is a military movie as well.

- Marine and Navy units participated in the filming of this movie, and after their work was finished they were sent to fight in Korea.

- This film was apparently 20th Century Fox's response to Warner Brothers' earlier Marine Corps war movie *Sands of Iwo Jima,* which was made a year earlier. Both movies were filmed at Camp Pendleton in California.

- Twentieth Century-Fox mogul Darryl F. Zanuck and producer Robert Bassler derived this movie's story from a short Marine training film called *Objective-Prisoners* which labeled this as "the key." The essence of the film's plot was the importance of capturing prisoners-of-war for interrogation as part of military intelligence efforts.

- Script treatments written by Sy Bartlett, Harry Kleiner and Major George A. Gilliland were not apparently used for this movie, although Gilliand did stay on as a credited technical consultant.

- According to a January 1949 *Los Angeles Times* news item, Dana Andrews, Anne Baxter and Paul Douglas were originally to star in the picture.

- Actual color combat footage shot in the war in the Pacific was incorporated into the film.

Reel Marines

- Proceeds from the film's premieres in New York and Los Angeles were donated to Marine Corps charities, and both were attended by Marine officials and veterans.

- The studio worked closely with the Corps to use the movie for recruitment, and a January 11, 1951 *Hollywood Reporter* news item noted a full company of recruits were sworn in at the film's San Francisco premiere.

- The W.E.B. Griffin novel *Under Fire* features the production of *Halls of Montezuma* as a part of the story.

- The landing craft are coming out of *USS Sedgwick County* (LST-1123), which saw action in WWII, Korea, and Vietnam.

- The movie's dedication, which is seen during the opening credits, states, "To the United States Marine Corps - this story is dedicated in gratitude for its help in making it possible - but most of all for its stalwart defense of all we hold dear to our lives, our people, and our future."

- Richard Boone, who would go on to star as Paladin in *Have Gun, Will Travel*, made his film debut in this movie.

- 20th Century-Fox announced this was the movie debut of Robert Wagner (as Private Coffman), however Wagner had previously appeared uncredited as Cleaves Catcher Adams in *The Happy Years* in 1950. Nonetheless, this movie still represents Wagner's credited movie debut.

Reel Marines

HEARTBREAK RIDGE

Release date: 5 December 1986
Running time: 130 minutes
Historical context: Grenada

Tagline: His past is Heartbreak Ridge. He is ready for another battlefield, and his finest hour. It will come.

<u>Cast</u>

Clint Eastwood - Gunnery Sergeant Thomas Highway
Mario Van Peebles - Corporal "Stitch" Jones
Arlen Dean Snyder - Sergeant Major Choozoo
Boyd Gaines - Lieutenant M.R. Ring
Everett McGill - Major Malcolm Powers
Moses Gunn - Staff Sergeant Luke Webster
Bo Svenson - Roy Jennings

Quote: "My name's Gunnery Sergeant Highway and I've drunk more beer and banged more quiff and pissed more blood and stomped more ass that all of you numb-nuts put together."
- Gunnery Sergeant Highway

Highlights

Get past some of the ridiculous aspects of this film, such as a group of Recon Marines acting like an undisciplined mob and treating a Medal of Honor recipient with contempt, and this is actually one of the most enjoyable films out there. Gunny Highway is the Marine we all want to be, and who among us has not known a Lieutenant Ring?

Reel Marines

Plot

Gunnery Sergeant Thomas Highway is nearing mandatory retirement from the Marine Corps and finagles a transfer back to his old unit. On the bus trip to his new assignment he meets fellow passenger "Stitch" Jones, a flashy wannabe rock musician who stiffs him for a meal at a stop and steals his bus ticket, leaving him stranded.

When Highway finally arrives at the base more bad news awaits. His new commanding officer, Major Malcolm Powers, is an Annapolis graduate who transferred over from Supply and has not yet had "the privilege" of combat. He sees Highway as an anachronism in the "new" Marine Corps, and assigns him to shape up the reconnaissance platoon. Recon is made up of undisciplined Marines who had been allowed to slack off by their previous platoon sergeant. Among his new charges, Highway finds none other than a very surprised and dismayed *Corporal* Stitch Jones.

Highway quickly takes charge and starts the men on a rigorous training program. They make a last-ditch attempt to intimidate him with the gigantic, heavily-muscled Marine named Swede Johanson (Peter Koch), but their plan fails miserably and they eventually begin to shape up and develop esprit de corps.

Highway repeatedly clashes with Powers and Staff Sergeant Webster over his unorthodox training methods, such as firing an AK-47 over his men's heads to familiarize them with the weapon's distinctive sound. Powers makes it clear that he views Highway's platoon as only a training tool for his own elite outfit, however Highway is supported by his old comrade-in-arms, Sergeant Major Choozoo, and his nominal superior officer, the awkward and inexperienced Lieutenant Ring. After Highway's men learn he had been

Reel Marines

awarded the Medal of Honor during the Korean War they gain renewed respect for him and close ranks against their perceived common enemy.

Highway also has his own personal problems. Aggie, his ex-wife, is working as a waitress in a local bar and dating the owner, Marine-hater Roy Jennings. Highway attempts to adapt his way of thinking enough to win Aggie back, even resorting to reading *Cosmopolitan* magazine to gain insights into the female mind. Aggie is initially bitter over their failed marriage, but tentatively reconciles with Highway – and then Highway's unit is activated for the invasion of Grenada.

Highway's platoon is dropped by helocast in advance of the main force. When confronted by a machinegun nest Highway improvises, ordering Stitch Jones to use a bulldozer to provide cover so they can advance and destroy the position. Next they rescue American students from a medical school and later, when they are trapped in a building by an armored car and infantry, radioman Profile (Tom Villard) is killed and his radio is destroyed. Lieutenant Ring shows previously unsuspected leadership qualities by coming up with the idea of using a telephone to call in air support.

Later, despite Powers' explicit orders to the contrary, the men take a key position - a historical fort. When Powers finds out he bawls them out and threatens Highway with a court-martial, but the commanding officer of the operation, Colonel Meyers, arrives and reprimands Powers for discouraging initiative and fighting spirit.

When Highway and his men return to the United States they are met by a warm reception, which is a first for Highway. Aggie is there to welcome him back, and to Highway's mock dismay Stitch informs him that he is going to stay in and make a career for himself in the Marine Corps.

Reel Marines

Trivia

❖ Clint Eastwood produced and directed *Heartbreak Ridge* in addition to starring as Gunny Highway.

❖ A portion of the movie was actually filmed on the island of Grenada.

❖ While the unit which participated in the actual invasion of Grenada and is portrayed in the film is the Camp Lejeune North Carolina-based 2nd Recon Battalion, the Quonset huts where filming took place are located in the Talega area of Camp Pendleton, California, which was at the time home to 1st Recon Battalion.

❖ The real Grenada invasion was codenamed Operation Urgent Fury and was carried out by the 22nd Marine Amphibious Unit, which was underway to Beirut at the time and was diverted enroute.

❖ October 25 was made a national holiday in Grenada, called Thanksgiving Day, to commemorate the invasion.

❖ The title comes from the Battle of Heartbreak Ridge in the Korean War. The character played by Eastwood was based on Private First Class Herbert K. Pililaau of Company C, 1st Battalion, 23d Infantry, who was posthumously awarded the Medal of Honor for his heroic actions there.

❖ Bo Svenson, who played Marine-baiting tavern owner Roy Jennings, actually served in the Marine Corps from 1959 to 1965.

Reel Marines

- The Marines planned to use the movie to promote their "Toys for Tots" campaign, but upon viewing a first cut quickly disowned the film because of the language.

- Marines who viewed the film cited numerous issues with how they were portrayed as being undisciplined and disrespectful. Highway's commanding officer is repeatedly shown disparaging and insulting him as well, while in reality this would have been extremely unlikely given his Medal of Honor.

- The sequence involving the bulldozer is based on a real event involving Army General John Abizaid, former commander of U.S. Central Command. During the invasion of Grenada the-Captain Abizaid improvised an attack on a Cuban bunker by having his unit take cover behind a moving bulldozer.

- The scene in which Lieutenant Ring must use a credit card in order to communicate with his commanders by telephone was loosely based on real-life events.

- "Stitch" Jones' title "Ayatollah of Rock and Rollah" is also used to describe the feudal warlord Humungus in the film *Mad Max 2*.

- When Clint Eastwood filmed this movie he was the Mayor of Carmel, California.

- The bar where "Little Mary" Jackson works is actually a place called "Carl's" in Vista, California, and is a favorite hangout of real Recon Marines.

- Alex M. Bello, who served with the author and retired years later as a First Sergeant, is credited as "Marine."

Reel Marines

- Prior to filming Mario Van Peebles could not play the guitar, but took several quick lessons to convince Clint Eastwood he could play the role of wannabe rock star Corporal "Stitch" Jones.

- Van Peebles wrote and performed the songs *Bionic Marine* and *Recon Rap.*

- Real Sergeant Major John H. Brewer appears as the Sergeant Major with Highway in court at the beginning of the movie.

- The battle of Heartbreak Ridge was actually fought mostly by the U.S. Army's 2nd Infantry Division. The battle became infamous after the Division Commander ordered the 23rd Infantry Regiment and an attached French infantry battalion to stage a disastrous frontal assault straight up Heartbreak Ridge. Sergeant Major Choozoo mentions he and Gunny Highway later joined the Marines after leaving the Army's 23rd Infantry Regiment.

- The Marine supervising Mario Van Peebles on the rappel tower is Staff Sergeant James Dever, who went on to retire as a Sergeant Major and founded "1 Force Inc.," a company which provides technical advisors to military themed productions.

- Screenwriter James Carabatsos, a Vietnam veteran of the 1st Cavalry Division, had a previous hit with his Vietnam War film *Hamburger Hill.* Inspired by an account of American paratroopers of the 82nd Airborne Division using a pay telephone and a credit card to call in fire support during the invasion of Grenada, Carabatsos fashioned a script of a Korean War veteran career Army

Reel Marines

non-commissioned officer taking his values to a new generation of soldiers. Clint Eastwood was interested in the script and asked his producer, Fritz Manes, to contact the Army with a view of filming the movie at Fort Bragg, however the Army read the script and refused to participate due to Highway being portrayed as a hard drinker, divorced from his wife, and using unapproved motivational methods - an image they did not want. The Army called the character a "stereotype" of World War II and Korean War attitudes that did not exist in the modern Army, and also did not like the obscene dialogue and lack of reference to women in the military. Eastwood pleaded his case to an Army general, contending that while the point of the film was that Highway was a throwback to a previous generation, there were values in the World War II and Korean War army which were worth emulating.

- ❖ Eastwood later approached the Marine Corps, which expressed some reservations about some bits of the film but provided support, so the character was changed to a Marine. This raised some conceptual difficulties, given that the Battle of Heartbreak Ridge primarily involved the Army, which was explained away when Sergeant Major Choozoo tells Corporal Jones that he and Highway were in the 2nd Infantry Division at the time and "joined the Corps later."

Reel Marines

INDEPENDENCE DAY

Release date: 2 July 1996
Running time: 145 minutes
Historical context: Future

Tagline: On July 2nd, they arrive. On July 3rd, they strike. On July 4th, we fight back.

Cast

Will Smith - Captain Steven Hiller
Bill Pullman - President Thomas J. Whitmore
Jeff Goldblum - David Levinson
Robert Loggia - General William Grey
Judd Hirsch - Julius Levinson
Randy Quaid - Russell Casse
Adam Baldwin - Major Mitchell
Harry Connick Jr. - Captain Jimmy Wilder

Quote: "Let's kick the tires and light the fires, big daddy!" - Captain Jimmy Wilder

Highlights

Some people may be surprised to find a science fiction movie on a list of Marine Corps films, but when the Chairman of the Joint Chiefs is a Marine General who is also the President's most trusted advisor, and the fighter pilot who saves the world is a Marine Captain, my question is, how can it NOT be?

Reel Marines

Plot

On July 2nd, an alien mothership with a mass equivalent to one-fourth of the Moon enters orbit around Earth and deploys several dozen saucer-shaped "destroyer" spacecraft, each fifteen miles in width, which position themselves over some of Earth's major cities. David Levinson, a cable company employee in New York City, discovers hidden satellite transmissions which he believes the aliens are using as a timer to coordinate a synchronized attack.

With the help of his ex-wife, a senior White House employee, David and his father Julius gain entrance into the Oval Office to warn President Thomas Whitmore of the impending attack. The President immediately orders large-scale evacuations of the targeted cities, but the aliens attack with advanced directed-energy weapons before these could occur. The President, his daughter, portions of his staff, and the Levinsons narrowly escape aboard Air Force One as the destroyers simultaneously lay waste to Washington D.C, New York City, Los Angeles, and several other major cities around the world.

On July 3rd, the United States conducts a coordinated counterattack. The Black Knights, a squadron of Marine Corps F/A-18 Hornets, participate in an assault on a destroyer near the remains of Los Angeles. Their weapons fail to penetrate the craft's force field, and it responds by releasing scores of smaller "attacker" ships which are similarly shielded and armed with directed-energy weapons, and a one-sided dogfight ensues. Marine Captain Steven Hiller survives by luring a single attacker to the Grand Canyon, where he blinds the alien with the braking parachute on his jet and ejects just before running out of fuel, causing both to crash in the desert.

Having parachuted to safety, he subdues the injured alien. Hiller is subsequently picked up by Russell Casse in the desert with a group of refugees in a convoy of RVs. They take the captured alien to nearby Area 51, where the President and his remaining staff have also landed. Here, it is shown Area 51 conceals a top secret facility housing a repaired attacker and three alien bodies recovered from Roswell in 1947.

When lead scientist Dr. Brackish Okun and his team attempt to remove a "bio-mechanical" suit from the alien, the specimen regains consciousness, attempts escape, and takes control of Okun's mind. When questioned by President Whitmore it reveals through a telepathic connection that its species travels from planet to planet, destroying all life and harvesting a planet's natural resources before moving on to the next planet. The alien attempts a psychic attack against Whitmore, but is killed. Whitmore orders a nuclear attack on all destroyers using B-2 Spirit bombers, but when a nuclear missile fails to penetrate the shield of a destroyer hovering over deserted Houston, Texas and destroys the city instead, the mission is aborted.

On July 4th Levinson devises a plan to use the repaired attacker to gain access to the interior of the alien mothership in space in order to introduce a computer virus and plant a nuclear bomb on board. This, it is hoped, will cause the shields of the Earth-based alien craft to fail long enough for the human resistance to eliminate them. Hiller volunteers to be the mission's pilot, with Levinson accompanying him to upload the virus. With satellite communications knocked out, the Americans use Morse code to coordinate an attack with the remaining forces around the world, timed to occur when the invaders' shields are set to fail. With few military pilots to man all available aircraft the battle requires several

Reel Marines

volunteers, including President Whitmore and Russell Casse, both of whom have previous combat flight experience.

With the successful implantation of the virus, President Whitmore leads the American jet fighters against an alien destroyer on approach to Area 51. Although the aliens now lack shields, the fighters' supply of missiles is quickly exhausted against the colossal craft and its large complement of attackers. The underside of the alien craft opens up as its weapon prepares to fire on the base. Casse possesses the last remaining missile, but his firing mechanism jams. He pilots his aircraft directly into the alien weapon in a suicide attack, and the ensuing explosion causes a chain reaction which annihilates the ship. Using this same method, human resistance forces around the world destroy the remainder of the alien ships while the nuclear device planted by Hiller and Levinson destroys the mothership soon after the duo escape. Hiller and Levinson return unharmed, crash-landing their captured attacker in the desert close to Area 51. The world celebrates, and the film ends as the main characters watch debris from the mothership enter the atmosphere like fireworks.

Trivia

- ❖ Captain Steven Hiller's ironic ambition before the alien attack is to join NASA's astronaut training program.

- ❖ Writers Dean Devlin and Roland Emmerich had always envisioned an African-American for the role of Hiller, and specifically wanted Will Smith after seeing his performance in *Six Degrees of Separation.*

- ❖ Not a single real fixed wing aircraft was in the air at any time in this film.

Reel Marines

- To prepare for the role of a former Persian Gulf War fighter pilot and current President of the United States, Bill Pullman read *The Commanders* by Bob Woodward and watched the documentary film *The War Room*.

- Marine Corps General William Grey is the Chairman of the Joint Chiefs of Staff and one of President Whitmore's most trusted advisors. Robert Loggia modeled the character after generals of World War II, particularly George S. Patton.

- The real Commandant of the Marine Corps from 1987 to 1991 was General Al Gray. The character of General William Grey may have been based upon him, since there are many similarities.

- Judd Hirsch's character of Julius Levinson was based on one of Dean Devlin's uncles.

- Adam Baldwin, who portrays Major Mitchell, was "Animal Mother" in *Full Metal Jacket*.

- Brent Spiner, best known as "Data" on *Star Trek: The Next Generation*, appears as Dr. Brackish Okun, the unkempt and highly excitable scientist in charge of research at Area 51. The character's appearance and verbal style are based upon visual effects supervisor Jeffrey A. Okun, whom Emmerich had worked with on *Stargate*.

- Singer Harry Connick, Jr. took over the part of Captain Jimmy Wilder for Matthew Perry, who was originally cast in the role.

- Singer Harry Belafonte had a cameo as one of the men looking at the spacecraft in New York City.

Reel Marines

❖ The United States military originally intended to provide personnel, vehicles, and costumes for the film, but backed out when the producers refused to remove Area 51 references from the script.

❖ Wendover Airport in Utah, formerly known as Wendover Air Force base, doubled for the El Toro and Area 51 exteriors. It was here that Pullman filmed his pre-battle speech.

❖ The President's speech was filmed on 6 August, 1995 in front of the old airplane hangar which once housed the *Enola Gay*, which had dropped the atomic bomb on Hiroshima exactly fifty years earlier on 6 August, 1945.

❖ The White House interior sets used had already been built for *The American President* and had previously been used for *Nixon*.

❖ The movie originally depicted Russell Casse being rejected as a volunteer for the July 4 aerial counter-offensive because of his alcoholism. He then uses a stolen missile tied to his red biplane to carry out his suicide mission.

❖ Smith's squadron was stationed at Marine Corps Air Station El Toro, a real air base in Orange County, California from 1943 until its decommissioning in 1999. The VMFA-314 "Black Knights," the squadron he Smith belongs to, was stationed at MCAS El Toro until 1994.

❖ Well over half of the dialogue in the scenes Jeff Goldblum shared with Judd Hirsch and Will Smith was improvised.

Reel Marines

- An entire scene in which Jeff Goldblum explains the nature of the alien signal had to be cut to avoid possible controversy due to a shot in which Harvey Fierstein planted an unscripted kiss on an unsuspecting Goldblum.

- The alien spacecraft "miniature" was over sixty-five feet wide.

- The White House which exploded was built at 1/12 scale, just to be blown up. Nine cameras filmed the explosion at various speeds, one of which was twelve times faster than normal and then played back at normal speed to make the explosion seem larger on film.

- In the briefing room scene at Area 51, behind Hiller and Grey, there is a night vision pan of the base. What you are seeing are actual shots of the real Area 51 taken by a conspiracy theorist from a place called "Freedom Ridge." The ridge was commandeered by the U.S. government in the late 90's and is no longer accessible to the public.

- The phrases said by the pilots when firing their missiles is NATO brevity code for the types of missiles being launched. "Fox One" means a semi-active radar-guided missile (AIM-7 Sparrow), "Fox Two" is an infrared-guided (heat-seeking) missile (AIM-9 Sidewinder), and "Fox Three" is an active radar-guided missile (AIM-120 AMRAAM).

- Spanish television advertisements for this movie showing the large ships hovering over New York were mistaken by some Spaniards for real disaster news footage - much as Orson Welles' *War of the Worlds* radio play had sparked an alien-war panic.

Reel Marines

- ❖ The scene in which Will Smith drags the unconscious alien across the desert was filmed on the salt flats near Great Salt Lake in Utah. Smith's line, "And what the hell is that smell?" was unscripted. Great Salt Lake is home to tiny crustaceans called brine shrimp. When they die, the bodies sink to the bottom of the lake (which isn't very deep) and decompose. When the wind kicks up just right the bottom mud is disturbed, and the smell of millions of decaying brine shrimp can be very bad - and apparently, nobody warned Will.

- ❖ The main helicopter used during the "Welcome Wagon" operation was a Sikorsky S-64 Skycrane which was outfitted with an array of flashing lights. When they first test-flew the helicopter with the lights on, over 150 calls were received in Orange County from people who had spotted the helicopter and, unsure of what it was, reported it as a "UFO sighting."

- ❖ When Will Smith enters the squadron locker room the extras watching television are real pilots from VMFAT-101, the Marine Corps' FA-18 Training Squadron.

Reel Marines

Release date: 4 November 2005
Running time: 123 minutes
Historical context: Gulf War

Tagline: Every man fights his own war.

Cast

Jake Gyllenhaal - Anthony Swofford
Scott MacDonald - D.I. Fitch
Peter Sarsgaard - Alan Troy
Jamie Foxx - Staff Sergeant Sykes

Quote: "You are no longer black, or brown, or yellow, or red! You are now green! You are light green! Or dark green! Do you understand?" – Drill Instructor Staff Sergeant Fitch

Highlights

While the book upon which this movie is based was clearly written by a disaffected "shitbird" who saw events from a very limited perspective, it is an important film because it is the only major motion picture about the Marine Corps' involvement in Operations Desert Shield and Desert Storm. We have all seen Marines do crazy things, and often their motivations are different from those of civilians, but the events which occur in Jarhead are the exception, rather than the rule, as the writers would have you believe.

Reel Marines

Plot

The film begins with voice-over narration on a black screen as Anthony Swofford (Jake Gyllenhaal) waxes philosophically about how a Marine's hands forever remember the grip of a rifle, whatever else they do in life. Swofford is then shown in Marine Corps boot camp, being brutalized by a drill instructor. After finishing recruit training "Swoff" is dispatched to Camp Pendleton, where a cruel joke is played on him by the senior Marines which involves branding onto him the initials "USMC" with a hot iron. He faints upon sight of the iron, and after regaining consciousness is told the hot iron had been switched the with another one which was room temperature. He is then greeted coolly by Troy (Peter Sarsgaard), who says to him, "Welcome to the Suck."

Swofford comes across the charismatic Staff Sergeant Sykes, a Marine "lifer" who invites Swofford to go through his Scout Sniper course. After arduous training sessions which claim the life of one Marine he becomes a sniper and is paired with Troy, who becomes Swofford's spotter. Shortly thereafter Kuwait is invaded by Iraq, and Swofford's unit is dispatched to the Persian Gulf as a part of Operation Desert Shield. Although the Marines are very eager to see some combat they are forced to hydrate, wait, patrol the nearby area, and orient themselves to the arid environment. When some field reporters appear Sykes directs his unit to demonstrate their NBC suits in a game of American football, even in the 112 °F heat. While the cameras roll the game develops into a rowdy dogpile, with some Marines playfully miming sex acts. Sykes, embarrassed by his platoon's rude manners and poor discipline, removes the cameras and crew from the area and the Marines are later punished by being

Reel Marines

made to build and take down a massive pyramid of sandbags on a rainy night.

During the long wait some of the Marines fear their wives and girlfriends at home will be unfaithful. A public board displays the photos of women who have ended their relationships with members of the unit, and Swofford himself begins to suspect that his girlfriend is now or will soon be unfaithful. The most public and humiliating of these befalls Dettman, who discovers an innocent looking copy of *The Deer Hunter* on a VHS tape sent by his wife is actually a homemade pornographic movie of her having sex with their neighbor, apparently as revenge for some unspecified act.

During an impromptu Christmas party Fergus, a member of Swofford's unit, accidentally sets fire to a tent and a crate of flares. Swofford gets the blame because he was supposed to be on watch but had Fergus sit in for him, and as a consequence Swofford is demoted from Lance Corporal to Private and made to undertake the degrading task of burning "diesel fuel" (excrement). The punishments, the heat and the boredom, combined with questions about his girlfriend's fidelity, temporarily drive Swofford to the point of mental breakdown and cause him to threaten and nearly shoot Fergus.

After the long stand in the desert the coalition forces' ground campaign, Operation Desert Storm, finally begins and the Marines are dispatched to the Saudi-Kuwaiti border. Just before the action begins Swofford learns from Sykes that Troy concealed his criminal record when enlisting, and will be discharged after the end of hostilities. Following an accidental air attack from friendly forces the Marines advance through the desert, but face no enemies on the ground. The units march through the Highway of Death, strewn with the burnt vehicles and charred bodies of

Reel Marines

retreating Iraqi soldiers, the product of the U.S. bombing campaign. Later the Marines encounter burning oil wells lit by retreating Iraqis, and they attempt to dig sleeping holes as a rain of crude oil falls from the sky. Before they can finish Sykes orders the squad to move to where the wind prevents the oil from raining on them, and while digging a new holes Swoff discovers Fowler has defiled an Iraqi corpse which drives him to the point of wanting to fight – but instead he takes the body and buries it somewhere else.

    Swofford and Troy are finally given a combat mission. Their orders are to shoot two Iraqi officers, supposedly located in a control tower at a battle-damaged airport. The two take up position in a deserted building, but moments after Swofford pinpoints one of the officers in his sights another team of Marines appears and calls in an air strike. Troy, desperate to make a kill, pleads with the officer in charge to let them take the shot. When his pleas are denied, Troy breaks down in a fit of despair and weeps, and moments later the airport is bombed by U.S. warplanes. Swofford and Troy linger at the site in a daze, losing track of time and missing their pick-up. With night falling they try to navigate the desert, but get lost. Distant cries in the darkness frighten them, and as they begin to sense the sounds are coming from beyond a ridge they ready their weapons and prepare to descend. They see an encampment in the distance, but on closer look recognize it as their base camp and the sounds as Marine voices. The war is over, they learn, and scores of Marines celebrate this amidst a bonfire. In a climactic scene Swofford tells Troy he never fired his rifle, getting a response of, "You can do it now." He then fires a round in the air from his sniper rifle, and the other Marines who also never had a chance to fire their weapons follow suit and empty their magazines into the night sky.

Reel Marines

On returning home the troops parade through towns in a jovial celebration of victory. The mood is disturbed when a disheveled Vietnam veteran, possibly suffering from his memories of the conflict, jumps onto their bus and congratulates them all. The veteran, still clearly disturbed by his experiences in Vietnam, asks the men if he can sit with them.

Soon after their return home Swofford and his comrades are discharged and go on with their separate lives. Swofford returns home to his girlfriend, but discovers her with a new boyfriend. Fowler is seen to be spending time with a girl at a bar, Kruger is shown in a corporate boardroom, Escobar works as a supermarket employee, Cortez becomes the father of three kids, and Sykes continues his service as a Master Sergeant in Operation Iraqi Freedom. An unspecified amount of time later, Swofford learns of Troy's death during a surprise visit from Fergus. He attends the funeral, meets some of his old friends, and afterwards he reminisces about the effects of the war.

Trivia

- ❖ Former Marine Captain Nathaniel Fick, the author of *One Bullet Away,* gave the film a mixed review (and panned the book on which it is based) in *Slate*. He wrote, "*Jarhead* presents wild scenes that probably could happen in combat units, but strips them of the context that might explain how they're more than sheer lunacy."

- ❖ The movie was filmed in the Imperial Valley of Southern California, which features conditions very similar to Iraq. Marines have in fact used one of the local towns, Brawley, for training purposes due to similarities to Iraq.

Reel Marines

- Most of Swofford's "anecdotes" are based on Urban Legends of the Marine Corps. He has made his unit the basis for, "Did you hear about that guy who..." for most USMC legends.

- One of the pictures on the "Wall of Shame," just to the left of center, is of porn star "Kitty."

- The word "fuck" and its variants are used 278 times in this film - thirty-eight times with the prefix "mother."

- Christian Bale, Leonardo DiCaprio, Tobey Maguire, and Josh Hartnett were considered for the role of Swofford.

- Michael Keaton, Kurt Russell, and Gary Oldman were all considered for the role of Lieutenant Colonel Kazinski.

- Jake Gyllenhaal's nosebleed during the prank branding scene was digitally added in post-production.

- Staff Sergeant Sykes, played by Jamie Foxx, originally had a tattoo of a panther on the back of his shaved head, and the actor sported it during his award sweeps for *Ray*. The tattoo was eventually digitally removed in post-production by director Sam Mendes because he felt it made the character too "hard core."

- While listed in the credits simply as "Swoff's sister," Jake Gyllenhaal's character refers to her as "Rini," which is in fact the real name of the actress who played her.

- Jake Gyllenhaal and Peter Sarsgaard actually became brothers in the real world, with the marriage of Sarsgaard to Gyllenhaal's sister Maggie.

Reel Marines

MAJOR PAYNE

Release date: 24 March 1995
Running time: 95 minutes
Historical context: Peacetime

Tagline: He's looking for a few good men... or a few guys old enough to shave.

<u>Cast</u>

Damon Wayans - Major Benson Winifred Payne
Karyn Parsons - Emily Walburn
Steven Martini - Cadet Alex Stone
Michael Ironside - Lieutenant Colonel Stone
Orlando Brown - Cadet Kevin "Tiger" Dunn
Albert Hall - General Decker
Geoffrey Kissell - Cadet Dotson
Damien Wayans - Cadet Dwight "D." Williams

Quote: "You want sympathy? Look in the dictionary between shit and syphilis." - Major Payne

Highlights

Major Payne may be a silly bit of escapist fare, but even Marines need some comic relief – and this film gives us a lot of opportunities to laugh at ourselves. I wonder how many of us nod our heads in complete understanding when Major Payne hangs upside down while field stripping his weapon blindfolded - more than a few, I would guess!

Reel Marines

Plot

Marine Major Benson Winifred Payne, a hardened killing machine, returns from a violent but successful drug raid in South America only to find out he was once again not promoted to Lieutenant Colonel. Payne receives an honorable discharge on the grounds that "the wars of the world are no longer fought on the battlefield," and that his military skills are no longer needed.

After he leaves the Marine Corps Payne finds life as a civilian unbearable and reaches his breaking point. To help adjust he applies for a job as a police officer, but during the test to see how applicants handle domestic violence disputes he overreacts and repeatedly slaps the man who hit his wife in the scenario. Payne is put into jail on charges of assault, and his former General visits him and informs Payne that he has secured him a job that will get him back in the military.

Payne arrives at Madison Preparatory Academy in Virginia, and is informed by the principal that his job is to train the Junior Reserve Officers' Training Corps "green boys," a disorderly group of delinquents and outcasts who have placed last in the Virginia Military Games eight years running. When Payne sees his company he immediately tells them that, under his direction, they will win the games at all costs. Regardless of their various shortcomings - being overweight, sickly, deaf, cross-eyed, orphaned, or from a dysfunctional home - they are all pushed equally. He also clashes with Emily Walburn, the Academy counselor who tries to soften Payne's discipline with understanding and feelings, particularly towards six-year old orphan Tiger.

Payne's training and punishments are harsh, which forces the cadets to attempt a series of failed schemes to get rid of him. Things come to a head when Payne offers them the

Reel Marines

chance to get rid of him, promising that if they can sneak into the rival school Wellington Academy and steal the Military Games trophy, he will leave voluntarily. He then places an anonymous call to Wellington, causing the boys to be ambushed by their rivals.

Payne bonds with Emily and Tiger, but when he returns to the academy he is confronted by lead misfit Alex Stone about his deception. Payne explains it was to show them what the real prize was, and with their desire to honestly earn the trophy added to their desire for revenge, and inspired by Payne saving Stone from his alcoholic stepfather, the boys begin to train hard to win.

Payne is later asked to return to the Marines to fight in Bosnia, but the deployment means he will miss the Military Games and disappoint the boys and Emily. As he waits for his train at the station he has a vision of himself, Emily, and Tiger barbecuing in a front yard, prompting him to realize that he has fallen for both Emily and the troops.

At the games the boys are holding their own until Dotson, one of the rival cadets (and the former squad leader who had transferred from Madison), purposely trips up Alex during a race and sparks an all-out brawl between the camps which threatens to get them disqualified. Payne, however, has given up his assignment, shows up at the last minute, and appoints Tiger to lead the cadence. The group then executes an unorthodox but entertaining drill routine which wins them the trophy.

On the first day of the new school year Payne resumes being an instructor, having settled down with Emily and Tiger, with Stone resuming his role as squad leader. When a new wise-cracking blind cadet shows up, Payne proceeds to shave him and his seeing-eye dog bald with his field knife, proving once again that he treats all recruits equally.

Reel Marines

Trivia

- ❖ The film is a very loose remake of the 1955 movie *The Private War of Major Benson* starring Charlton Heston.

- ❖ While generally receiving unfavorable reviews, the movie has become a cult favorite, especially with JROTC/ROTC/Civil Air Patrol cadet programs.

- ❖ In one scene Major Payne tells the children, "What we have here is a failure to communicate!" This is a famous quote taken from the movie *Cool Hand Luke.*

- ❖ The theme song playing when Major Payne is dreaming about being married with the picket fence is from 1958's *The Donna Reed Show.* The husband on that show was Alex Stone, which is also the name of the cadet who leads the unit.

- ❖ Payne refers to the cadets as "soldiers." A real Marine would *never* make that mistake!

- ❖ On several occasions Payne salutes General Decker while indoors and uncovered. Once again, no real-life Marine would have done that.

- ❖ In the opening scene of the movie after the raid in South America a sign is shown which reads "Camp Pendleton, Virginia." That is a state-owned military reservation for the Virginia National Guard, not the Marines. The Camp Pendleton the Marine Corps occupies is located in Southern California, near San Diego.

- ❖ Damien Wayans, who plays deaf Cadet Dwight "D." Williams, is the nephew of Dwayne Wayans, Keenen Ivory Wayans, and Damon Wayans.

Reel Marines

PRIDE OF THE MARINES

Release date: 24 August 1945
Running time: 119 minutes
Historical context: World War II

Tagline: Based on a true story, a Marine falls in love and is blinded fighting the Japanese at Guadalcanal, but has difficulty adjusting to his changed life.

Cast

John Garfield - Al Schmid
Eleanor Parker - Ruth Hartley
Dane Clark - Lee Diamond
Anthony Caruso - Johnny Rivers
Rosemary DeCamp - Virginia Pfeiffer

Quote: "Now that I'm going home, I'm scared. I wasn't half as scared on Guadalcanal as I am now." - Al Schmid

Highlights

The Battle of the Tenaru River depicted in this film goes a long way toward explaining the "fog of war" to those who have never experienced it. As good as the combat scenes are, the real movie is a character study in which a real-life hero must come to grips with his scars, both physical and emotional. Pride of the Marines has become relevant once again, as more and more wounded veterans return from Iraq and Afghanistan.

Reel Marines

Plot

The film is divided in three parts. The first takes place prior to the war, where cocky Philadelphia steel worker Al Schmid resists the idea of marriage and losing his independence until he meets his match in Ruth Hartley.

In part two, at the Battle of the Tenaru River on Guadalcanal, Schmid is in the crew of an M1917 Browning machinegun as part of H Company, 2nd Battalion, Fifth Marines with his buddies, Jewish Lee Diamond and American Indian Johnny Rivers. While the three wait for an enemy attack they practice gun emplacement procedures, establish fields of fire, practice with the range card to estimate firing distances, and determine optimal traversal and elevation settings for each anticipated line of attack.

The subsequent onslaught is heavy. Rivers is killed by a bullet through the head, Diamond is wounded by three bullets in his right arm, and Schmid is blinded when a Japanese soldier drops a hand grenade at the front of the gun pit. In spite of the heavy attack the sightless Schmid is still able to fire his weapon by following Diamond's instructions, and together they kill two hundred of the enemy.

The third part of the movie follows Schmid's humbling rehabilitation in which he resents being dependent upon others. He hopes that an operation will restore his sight, but the medical procedure wasn't successful. He doesn't want Ruth to know that he is almost completely blind, and attempts to break up with her, but eventually learns responsibility through Diamond, hospital rehabilitation officer Virginia Pfeiffer, and the other wounded veterans.

He is to be awarded the Navy Cross, but is dismayed when he finds out the ceremony will take place in his hometown. He initially feels anger and discomfort when he becomes

Reel Marines

dependent upon family and friends, primarily because he doesn't want to be a burden to anyone, but in spite of his resentment Ruth stays by his side and helps him overcome his bitterness and convinces him he must learn to live with his disability.

Trivia

- During the Battle of Guadalcanal two enlisted Marines, Mitchell Paige and John Basilone, were awarded the Medal of Honor for their use of the M1917 Browning machine gun against massed Japanese charges.

- In Jim Proser's book *I'm Staying With My Boys: The Heroic Life of Sgt. John Basilone USMC* Proser tells of Basilone's friendship with star John Garfield when he toured the United States selling war bonds.

- Screenwriters A. I. Bezzerides and Alvah Bessie developed a twenty-six page treatment of Roger Butterfield's book *Al Schmid Marine*. Martin Borowsky also did an adaptation of Butterfield's book that was rewritten by Albert Maltz - whom Garfield had spoken to about Butterfield's story.

- Prior to filming, Garfield visited American soldiers in hospitals in Italy.

- Garfield met Schmid during his rehabilitation before a movie was ever envisioned.

- Once the film was planned, Garfield lived with the Schmids for several weeks and became friends with the couple.

Reel Marines

PURPLE HEARTS

Release date: 31 August 1984
Running time: 116 min
Historical context: Vietnam

Tagline: Nothing could have prepared him for the danger, the fear, the violence... or the woman.

Cast

Ken Wahl - Lieutenant (USN) Don Jardian
Cheryl Ladd - Lieutenant (USN) Deborah Solomon
R. Lee Ermey - Gunny
Stephen Lee - Wizard
James Whitmore Jr. - Bwana

Quote: "Marines don't die on Army choppers. Don't give those doggie pricks the satisfaction!" - Gunny

Highlights

Purple Hearts is the rare sort of movie that a couple can agree to watch together, because she will enjoy the sappy love story, and he will love the combat sequences. Any way you look at it, R. Lee Ermey's short but memorable turn as a salty platoon sergeant is worth the price of admission, and the scenes where the firebase is being overrun are far better than one might expect in a film of this sort. The bottom line is while Purple Hearts isn't in the same class with Sands of Iwo Jima, it's still quite watchable.

Plot

Purple Hearts flings together Navy Dr. Don Jardian and Nurse Deborah Solomon against the backdrop of war-torn Vietnam. The film begins with a dedication to the Vietnam War's 347,304 Purple Heart recipients, though it also seems aimed at the nation's millions of soap opera fans. Whenever the film's war scenes verge on becoming believable, which is often enough, the love story is allowed to take over.

Before we ever meet the main characters the movie opens with a platoon of Marines being inserted into a hot landing zone by helicopter. The platoon sergeant, played by R. Lee Ermey, gets to speak most of the film's best lines, and when one of his Marines is wounded and medevaced to a Navy field hospital we are introduced to Jardian, the "Wizard," and the rest of the medical staff.

When we first meet Dr. Jardian he is speaking fluent doctorese ("Prep him for O.R., this arm and leg gotta come off right now!") and is frantically overworked. Somehow, he then abandons all of his other patients to fly this particular wounded Marine to Da Nang. There he is watching though the operating room window while a team of surgeons attempts an emergency operation when he spots a pair of pretty eyes above one particular surgical mask. That's Nurse Solomon, and since they are stationed at different bases the pair spend the rest of the movie trying to arrange free weekends in order to visit one another, just as they might in a college romance.

The pivotal moment comes when Jardian goes too far out on a limb to see Solomon, and as punishment he is sent to a remote Marine firebase where he learns about war the hard way before later participating in a secret POW rescue mission - and becomes a better man for it.

Reel Marines

Trivia

❖ The Purple Heart is of course a U.S. military decoration for wounds received in battle, and the title is a coy way of juxtaposing the combat elements of the film with the underlying love story.

❖ *Purple Hearts* was filmed in the Philippines.

❖ R. Lee Ermey's character, credited only as "Gunny," tells his new platoon commander, "That's the second goddam thing you did right today, sir!" when they enter a hot LZ as the movie opens. Many NCO's who have broken in a new officer love this scene.

❖ Star Cheryl Ladd was once the daughter-in-law of Alan Ladd, who was most famous as *Shane*, and played a Marine in the 1960 film *All the Young Men.*

❖ James Whitmore Jr., who plays "Bwana," played a Marine fighter pilot in *Black Sheep Squadron* and is the son of actor and Marine Corps veteran James Whitmore.

Reel Marines

RETREAT, HELL

Release date: 6 October 1952
Running time: 94 minutes
Historical context: Korea

Tagline: A bunch of husky guys in battle-green who showed the world you can't stop a Marine.

Cast

Frank Lovejoy - Lieutenant Colonel Steve L. Corbett
Richard Carlson - Captain Paul Hansen
Nedrick Young - Sergeant Novak
Lamont Johnson - Captain "Tink" O'Grady
Robert Ellis - Shorty Devine
Peter Ortiz - Major Knox
Russ Tamblyn - Jimmy W. McDermid

Quote: "Retreat, Hell!" – General O. P. Smith

Highlights

The 1st Marine Division's landing at Inchon and breakout from the Chosen Reservoir are two of the most legendary events in the long and proud history of the Marine Corps, and any movie about them would be a must-see even if it wasn't very good. Fortunately, that was not the case with Retreat, Hell! This movie tells the story well from a historical perspective, while weaving in the compelling personal stories of the Marines from beginning to end.

Reel Marines

Plot

Retreat, Hell! is a 1952 film about the 1st Marine Division in the Korean War. It follows the war from the formation of the 1st Marine Division, through the historic landing at Inchon and the subsequent capture at Seoul, and concludes by depicting the epic, fighting withdrawal from the Chosen Reservoir. The story ends there, because the Korean conflict would not end until almost a year after the release of this motion picture.

In 1950 top U.S. military officials are ordered to send every available unit to Korea, and after a Marine division is quickly put together from spare units and reserve personnel Paul Hanson, a Reserve captain and World War II veteran, is ordered to Camp Pendleton in Southern California. As he drives onto the base with his wife and two young daughters he confides to her that he feels much older than the other Marines and far from his military training.

More disappointment follows when Hanson learns from stern Colonel Corbett that he is to lead an infantry company at the front, an assignment which is unrelated to his electronics training. The next day he and the men are sent into the hills to begin intensive training, and not everyone is as reluctant as Hanson. Young Jimmy McDermid, the brother of a Marine already fighting in Korea and another killed at Iwo Jima, is anxious to prove himself. So anxious, in fact, that he lied about his age in order to enlist.

After many grueling days of mock combat in the California hills the men return to camp, the officers spend one last night with their families, and the next morning the division is shipped out. Aboard ship Corbett admits that he has little faith in Hanson, a "retread captain," whom he expects "will play it too safe."

When they reach Korea, the division makes an amphibious landing at Inchon and is immediately engaged in battle. When a Marine is killed in front of him Jimmy's eagerness turns to paralyzed fright. Lagging behind, he reaches the bivouac area after the others and is too ashamed to meet with his brother, who is in a nearby recon unit.

The next day the Marines fight their way toward Seoul, and once in the city they are held at bay by several enemy troops, including one with a machinegun firing from a second floor window. Jimmy spots him, and works his way cautiously through the gunfire to toss a grenade into the window.

With the situation under control Corbett orders a break and, seeing that Hanson has received his first letter from home, intimates his regrets at having no family. Hanson responds by pointing out that the men of the division are his family. Then, when Corbett and Hanson ride to headquarters to receive their orders, an exuberant Jimmy comes along to see his brother - but upon arrival finds out that he has just been killed.

Despite a rumor that they will be home by Christmas, Corbett is ordered to push through Korea all the way to the Chinese border. The Marines move forward and walk into an ambush, but after a fight they destroy an enemy tank. Jimmy performs well during the battle, but Corbett sees that after his brother's death his innocent determination to be a good Marine has changed into a grim war fever. As a reward, but also out of concern, Corbett reassigns Jimmy to replace his wounded driver.

Later, after spotting an elusive enemy unit which is shadowing their movements, Corbett is surprised to realize that the Chinese have joined the North Koreans. Then in early December he announces their orders to travel through a

Reel Marines

narrow mountain road to get to the northern border. He has also been ordered to send Jimmy home, as he is now the sole surviving son in his family.

Unhappy with the news, Jimmy asks to spend his last night with friends in Hanson's company, but during the night they are attacked by enemy soldiers who break through one of the company lines. Corbett radios for help, but is told there is no one to spare and that supplies cannot be dropped until it is light enough for the pilots to see. When the men run out of ammunition they prepare to fight with bayonets, but supplies arrive in time for them to fend off their attackers.

After Corbett conveys new orders that they are to retrace their steps and head toward the sea, the men ask if they are retreating. Corbett snaps, "Retreat, Hell! We're advancing in a different direction!" and promises that he will not leave behind the wounded.

Although they successfully squelch another major attack, the winter storms bring new dangers. Hanson enlists a group of volunteers to go into the hills to rescue the wounded, and gets unexpected help from a unit of British Royal Marines - but the rescuers are ambushed and lose contact with the main column.

Meanwhile, on the road, the Marines suffer another attack and Corbett is shot. When he regains consciousness he is informed that the division made it to an airstrip, but that Hanson's men and the British are lost. As Corbett despairs, Hanson, Jimmy and the others march up. Relieved and confident, Corbett announces that they will fight their way to the sea, and the First Marine Division pushes on.

Reel Marines

Trivia

❖ "Retreat, Hell" is the motto of the 2nd Battalion, 5th Marine Regiment.

❖ Peter Julien Ortiz, who appeared in the film as Major Knox, was a highly decorated Marine who served in the Office of Strategic Services (OSS) during World War II and later appeared in several films.

❖ *Variety* called *Retreat, Hell!* a "top-notch war drama" for the way it balanced tense action with a more human face of war and for anticipating film-making trends which would become more common twenty years later.

❖ With the 1st Marine Division's fight for life at the Chosin Reservoir against the Chinese Communist Forces in the winter of 1950 being anxiously followed in the news of the day, Warner Brothers submitted a proposal to the Marine Corps to make a film about the events. The Corps approved the request, with former Marine Milton Sperling producing and co-writing the film for the United States Pictures division of Warner Brothers.

❖ The film crew bulldozed a road at Camp Pendleton and sprinkled the area with gypsum to simulate snow. The Marines also created accurate Korean villages for the film.

❖ The Hollywood Production Code Office originally refused to approve the title because of its ban on the word "hell," but changed their mind after a direct request from the Marine Corps.

Reel Marines

❖ The title is based on a defiant quote from General Oliver P. Smith which gained fame during the Korean War.

❖ The Korean War was the first war where United States troops were desegregated. An oral history interview with Donald H. Eaton, a black Korean War veteran, includes a story where he related how he and several friends watched this film when it came out and half of his friends ended up enlisting in the Marine Corps.

Reel Marines

RULES OF ENGAGEMENT

Release date: 31 March 2000
Running time: 128 minutes
Historical context: Vietnam/Peacetime

Tagline: A hero should never have to stand alone.

<u>Cast</u>

Tommy Lee Jones - Colonel Hayes L. "Hodge" Hodges II
Samuel L. Jackson - Colonel Terry L. Childers
Guy Pearce - Major Mark Biggs
Ben Kingsley - Ambassador Mourain
Bruce Greenwood - National Security Advisor Bill Sokal
Blair Underwood - Captain Lee
Dale Dye - Major General Perry

Quote: "Have you ever had a pissed off Marine on your ass?"
 - Colonel Hodges

Highlights

Anyone who has ever had to deal with the rules of engagement which are all too often dictated to Marines serving in harm's way by diplomats who are safely back in the rear will appreciate the message of this movie. Combat is like a big, dirty knife fight, and the words of big Harvey in Butch Cassidy come to mind - "Rules? In a knife fight?" Far too often we are asked to observe the Marquis of Queensbury Rules, while the other guys get to fight dirty!

Reel Marines

Plot

The film opens during the Vietnam War in 1968 near Ca Lu. A Marine platoon led by Lieutenants Hays Hodges and Terry Childers is advancing through the jungle, and they split up to take two different routes. Hodges' group is ambushed by North Vietnamese soldiers and all but Hodges are killed. At the same time Childers' group captures the leader of the attackers, Colonel Binh Le Cao. Childers orders Colonel Cao to call his men off Hodges, and to try and intimidate him he holds a pistol to Cao's unarmed radioman and promises to let them both go free if they cooperate. When Cao refuses to call his men off Childers executes the radioman and turns the gun on Cao, who immediately radios his men and tells them to let Hodges live. Childers then keeps his word and lets Colonel Cao go.

The movie then jumps to twenty-eight years later. Hodges is now a Colonel, although he has been confined to a desk job since the war years due to an injury picked up in Vietnam. He is having a retirement party, and as a surprise Colonel Childers shows up to present him with a sword.

Not long after this Childers and his Marines are called into action in Yemen where an unruly crowd of local men, women and children demonstrate outside the U.S. Embassy in Sana'a after being incited by jihad audio tapes. The U.S. Ambassador, along with his wife and young son, needs to be evacuated, and during the operation three Marines are killed by gunmen on the roofs of buildings overlooking the Embassy.

After retrieving the U.S. flag and flying Ambassador Mourain and his family out of the Embassy, Childers returns to aid his Marines. A fourth Marine, Sergeant Kresovitch, is then mortally wounded and dies in Childers' arms, and moments later the Colonel appears to see something in the crowd below and immediately orders his men to open fire on the crowd, killing eighty-three men, women and children.

In the wake of this incident U.S. National Security officials worry about the dire consequences of Childers' actions and decide he needs to be court-martialed. The legal case which follows is dependent upon whether the crowd was armed and fired first, or Colonel Childers exceeded his orders and reacted based on anger, confusion or a darker motive such as prejudice. Childers faces three charges – murder, for killing eighty-three "non-combatants," conduct unbecoming an officer and a gentleman, and the minor charge of breach of the peace.

If some of the people in the crowd were carrying weapons and opened fire Childers could be exonerated, so he asks Hays Hodges to become his attorney. Hodges tells him he needs a better lawyer than him in order to avoid a life sentence because Hodges had an unimpressive career in the JAG Division, but Childers is adamant about having him as his attorney because he has served in combat before. Hodges agrees and goes to Yemen, where he finds several audio tapes which call for a jihad against the USA and explain the mob outside the Embassy and the shooting.

The prosecution asserts that Childers' order to fire was based on personal fear, racism, or confusion, and National Security Advisor William Sokal wants him to be convicted in order to preserve U.S. relations with Arab countries. When Sokal receives a CCTV security videotape which clearly shows hostile fire coming from the crowd outside the Embassy, he burns it in the hope the prosecution will win. The defense and Childers respond that he was in fear for his Marines' lives under fire, and was in compliance with his orders and the rules of engagement. Childers testifies that he was on the roof and could clearly see that the crowd had weapons, but another Marine who also had a good observational position was killed on site – leaving Childers as the only one who can testify as to the intentions of the crowd. Ambassador Mourain, who has been blackmailed into cooperating by Sokal, lies on the stand and says the crowd was peacefully demonstrating, and adds that Childers had acted violently towards him during the evacuation. His wife later admits the truth to Hodges, but won't testify and contradict her husband. The prosecution then introduces previous actions by then-Lieutenant Childers in Vietnam to show a history of misconduct, including a witness named Colonel Binh Le Cao - the very man Childers had captured and released in Vietnam.

Colonel Cao recounts how Childers threatened him with death to save his Marines and executed his unarmed radioman, but admits that if placed in the same situation, he would have done the same thing. Hodges also presents the jury with a shipping manifest showing that film from an undamaged camera which had the exact same point of view as Childers had been delivered to Sokal's office but had since disappeared. This is a turning point in the trial, and ultimately Childers is found guilty of breach of the peace but not guilty of the more serious charges. In an effort to save face the attorney for the prosecution tries to get Hodges to recount what he remembered about the incident in Vietnam to pursue this against Childers, but Hodges responds by reminding the Major of his lack of combat experience and tells him the life expectancy of a Marine dropped into a hot LZ was not one week as Biggs had thought, but rather "sixteen minutes." As Childers steps outside the courthouse Colonel Cao salutes him before getting into his car, and Childers salutes back.

Reel Marines

Trivia

❖ Lead writer James H. Webb is a decorated former Marine combat officer and Secretary of the Navy, and is currently a United States Senator from Virginia.

❖ The film drew widespread criticism for its portrayal of Arab characters, with the American-Arab Anti-Discrimination Committee describing it as "probably the most racist film ever made against Arabs by Hollywood." Paul Clinton of the *Boston Globe* wrote "at its worst, it's blatantly racist, using Arabs as cartoon-cutout bad guys." 9-11 took place the following year, and rendered those opinions moot.

❖ *USS Wake Island* (LHA-7) is fictional and not an actual Navy ship. The ship seen in the movie is actually *USS Tarawa* (LHA-1), although the fictional *Wake Island* is often mentioned in the television military drama "JAG."

❖ A replica of the Marine Corps Base at Camp Lejeune, North Carolina, was built at the Vint Hill Farm Station in South Manassas, Virginia for this movie.

❖ Eight hundred Marines were hired as extras for the combat and other action scenes.

❖ When Hodges (Tommy Lee Jones) returns to the bombed-out embassy there is a picture of then-Vice President Al Gore on the charred wall. Gore and Jones were once roommates at Harvard.

❖ At the beginning of the movie some Colt M16 VNs can be seen being used by the platoon of Marines, along with a few M60 machine guns and a Remington 870 Combat

Reel Marines

Shotgun, plus a couple of RPD machine guns and some AK-47s in the hands of the Vietcong. Later on the Marines arriving at the Embassy carried M16A2s and M249 SAWs, and the Embassy Guards were armed with Mosberg Shotguns.

Reel Marines

SALUTE TO THE MARINES

Release date: September 1943
Running time: 101 minutes
Historical context: World War II

Tagline: They've never been licked!

Cast

Wallace Beery - Sergeant Major William Bailey
Reginald Owen - Henry Caspar
Ray Collins - Colonel John Mason
William Lundigan - Rufus Cleveland
Noah Beery, Sr. - Adjutant
Dick Curtis - Corporal Moseley
Marilyn Maxwell - Helen Bailey
Hugh Beaumont - Sergeant (uncredited)
Robert Blake - Junior Carson (uncredited)
Jim Davis - Private Saunders (uncredited)

Quote: "I'm the only non-combatant Marine in the history of the Marine Corps!" - Sergeant Major Bailey

Highlights

Major Gene "Dunk" Duncan once wrote that any Marine who does not yearn to be in the middle of a raging battle is all image, and no substance – and because of that the character of Sergeant Major Bailey is one with which many Marines who have never gone to war can identify.

Reel Marines

Plot

William Bailey, a tough but dedicated Sergeant Major stationed in the Philippines, is the friend and trainer of "Flashy" Logaz, a former lightweight world champion who hopes to regain his title. One day his Commanding Officer, Colonel Mason, sends for Bailey and introduces him to a member of the Philippine government, which is due to receive its independence in 1945. Mason tells Bailey he is to train Philippine civilians but Bailey, who is soon to retire, hates the thought and wants to see action instead. He is at first frustrated by recruits who do not understand military discipline and prefer traditional bolas to bayonets, but a month later the well-trained recruits have earned his respect.

When Bailey hears that his battalion is being sent to Shanghai he rushes to Mason, who had promised him a transfer if the unit was going to see action, but the Colonel says neither of them will be going because of their age. That night Bailey is arrested by MPs after he goes on a binge and gets into a fight with some merchant marines. After Bailey's wife and daughter visit him in the brig, they successfully pleads with Mason to have him discharged early.

Bailey then goes to live with his family in the nearby town of Balanga, where many American and British ex-patriots reside. Bailey is uncomfortable with civilian life and has a hard time adjusting, so he starts to train the village children as he continues to coach Flashy. The village mothers do not want their children doing military training and Bailey's daughter Helen, who neither believes in war nor understands his frustration, is disgusted.

When some Japanese merchant marines arrive with supplies Casper, the storeowner, tells the suspicious Bailey the men are "peace-loving," but Bailey and Flashy think their ship is too well equipped. As Christmas approaches Jennie and Bailey have become closer, and she is hopeful peace talks in Washington will go well until one Sunday morning Rufe Cleveland, who is a Marine pilot, encounters Japanese planes while flying and gets the message that Pearl Harbor has been bombed. At the same time villagers hear planes as they attend church, and soon bombs begin to fall.

After the attack Casper urges the villagers to fight against American forces, but Bailey comes into the crowd carrying the body of a dead child and reveals Casper for what he is, a Nazi who has been secretly aiding the Japanese. He then rallies the villagers by saying Filipinos are just as American as he is, and kills Casper with his bare hands.

Just as Bailey starts to organize the villagers, Rufe arrives. He wants to help, but Bailey insists he fly to headquarters to warn them, while taking the women and children with him. A short time later sailors from the Japanese ship don their military uniforms and invade the village. Bailey directs the battle, and leads the way toward a bridge which is critical to the Japanese. As the battle rages Bailey proudly sees that Jennie has stayed behind to help the wounded, and soon increasing numbers of Japanese foot soldiers, tanks and planes attack as American reinforcements arrive.

After a large loss of life, the Americans successfully dynamite the bridge and rout the Japanese. Bailey then tells the men to take to the hills and fight the way they know best, because the Japanese are going to try to take over the islands. After sadly bidding Flashy and the other Filipinos goodbye, Bailey and Jennie embrace as the Japanese bombing resumes. Later, back in San Diego, Helen - who is now a member of the women's reserve of the Marine Corps - accepts her father's posthumous Medal of Honor and says goodbye to Rufe as he leaves for a new post.

Reel Marines

Trivia

- Released on the heels of the attack on Pearl Harbor, *Salute to the Marines* pulsates with the patriotism of its era, and also includes the expected cultural stereotypes and derogatory depiction of the enemy.

- The film was viewed by many as an allegory for the U.S. entry into WWII because Bailey had never seen combat until his home was attacked, and he stepped up to do his duty.

- Bailey's daughter Helen begins the movie as a pacifist, but in the end a tragic event makes her understand the sacrifice of war, just as it took Pearl Harbor to motivate America to enter World War II.

- Wallace Beery's brother Noah Beery, Sr. appears in *Salute to the Marines* as a Marine Adjutant.

- Noah Beery died three years after the film's release in 1946 of a heart attack in Beverly Hills on his brother Wallace's birthday.

- Noah Beery's son Noah Beery Jr. appeared as a Marine in *Gung Ho!* the same year his father made *Salute to the Marines*.

- Oscar winner Fay Bainter starred in the supporting role of Sergeant Major Bailey's wife, Jennie. The actress made Oscar history in 1938, when she was nominated for two Academy Awards in different categories.

- Future blonde bombshell Marilyn Maxwell received her first substantial part in *Salute to the Marines* as the Bailey's idealistic daughter.

Reel Marines

- Maxwell was a crowd-pleaser and stalwart trouper in Bob Hope's legendary tours of military installations around the world, and part of her USO touring act with Hope was wearing a tight sweater and singing *I Want to Love You*.

- Hugh Beaumont, who would one day play father Ward Cleaver on *Leave it to Beaver*, appears uncredited as a "Sergeant."

- Ten-year-old Robert Blake, who gained fame years later as TV's *Baretta* and more recently was tried for murdering his wife, turns up in *Salute to the Marines* credited as Bobby Blake and plays a character named Junior Carson.

- The film is unique in that it has a "light comedy" aspect to it, with Wallace Beery managing something like an Archie Bunker quality - probably much needed comic relief, given the news from the front at the time.

- William Lundigan, who appears as Rufus Cleveland, joined the Marine Corps the same year this movie was released and fought on Pelileu with 3rd Battalion, 1st Marines and later participated in the invasion of Okinawa.

- Jim Davis, whose greatest fame would come almost forty years later as patriarch Jock Ewing in TVs's *Dallas*, has an uncredited part as Private Saunders.

- *Salute to the Marines* premiered at Chico's Senator Theater, where it raised $300,000 in war bonds.

Reel Marines

SANDS OF IWO JIMA

Release date: 14 December 1949
Running time: 100 minutes
Historical context: World War II

Tagline: Alone and outnumbered, they had one thing in their favor... the American dream.

Cast

John Wayne - Sergeant John M. Stryker
John Agar - PFC Peter Conway
Forrest Tucker - PFC Al Thomas
Wally Cassell - PFC Benny Regazzi
James Brown - PFC Charlie Bass
Richard Webb - PFC Dan Shipley
Arthur Franz - Corporal Robert Dunne / Narrator
Hal Baylor - PFC "Ski" Choynski

Quote: "Life is tough. It's tougher when you're stupid."
 - Sergeant Stryker

Highlights

Sands of Iwo Jima is the gold standard against which all war films are measured. It has the biggest star in John Wayne, the most iconic battle in Iwo Jima, and even features some real Marine heroes. Let's face it, every man who has ever worn the Eagle, Globe and Anchor wants to be Sergeant Stryker - whether he admits it or not!

Reel Marines

Plot

Tough-as-nails career Marine Sergeant John Stryker is greatly disliked by the men of his squad, particularly the combat replacements, because of the rigorous training he puts them through. He is especially despised by Private Peter Conway, the arrogant, college-educated son of an officer Stryker served under and admired, and Private Al Thomas, who blames him for his demotion.

When Stryker leads his squad in the invasion of Tarawa the men begin to appreciate his methods - except for Conway, who considers him brutal and unfeeling when he apparently abandons a wounded comrade to the enemy. During the battle Thomas goofs off when he goes to get ammunition for two buddies, stopping to savor a cup of coffee, and as a result he returns too late and the two Marines, now out of ammunition, are overrun. Hellenpolis is killed, and Bass is badly wounded, and when Stryker discovers the truth he forces Thomas into a fistfight. This is seen by a passing officer, but Thomas unexpectedly gets Stryker out of trouble for hitting a subordinate by claiming he was being taught judo. His conscience is bothering him, and Thomas breaks down and abjectly apologizes for his dereliction.

Stryker shows his soft side while on leave in Honolulu when he picks up a bargirl and goes to her apartment. He becomes suspicious when he hears somebody in the next room, but when he investigates all he finds is a hungry baby boy she is supporting the best way she can. He gives the woman, who is the widow of a Marine, all of his money and leaves.

Later, during a training exercise, a Marine drops a live hand grenade and everybody drops to the ground except

Reel Marines

Conway, who is distracted reading a letter from his wife. Stryker knocks him down, saving his life, and then proceeds to bawl him out.

Stryker's squad next fights in the battle for Iwo Jima, and witnesses the iconic flag raising on Mount Suribachi. Afterwards, while the men are resting during a lull in the fighting, Stryker is killed by a sniper. His men find a letter on him addressed to his son which says all of the things he wanted to say, but never got around to.

Trivia

❖ The flag used in the movie was the actual one raised on Mount Suribachi during the battle. It was on loan from the Marine Corps Museum in Quantico, Virginia.

❖ The title *Sands of Iwo Jima* was once seen by the movie's producer Edmund Grainger in a newspaper story. Grainger thought of the American flag raising at Mount Suribachi, and went off and wrote a treatment with this title and ending with the famous flag raising scene.

❖ Grainger wanted the movie to have an influence on the public's attitude towards the Marines at a time when the Corps was fighting with Congress for survival and needed more financing.

❖ Appearing as themselves are Lieutenant Harold Schrier, who led the flag-raising patrol on Iwo Jima, Colonel David M. Shoup, later Commandant of the Marine Corps and recipient of the Medal of Honor at Tarawa, and Lieutenant Colonel Henry P. "Jim" Crowe, commander of the 2nd Battalion 8th Marines at Tarawa, where he earned the Navy Cross.

Reel Marines

- Lieutenant General Holland M. "Howlin' Mad" Smith, who was the 5th Amphibious Corps' Commander, acted as a technical advisor to the film.

- The film has the first recorded use of the phrase "lock and load," twice as a metaphor for "load your weapons and get ready to fight," and once as an invitation to drink alcohol (get loaded).

- During World War II Arthur Franz, who played Dunne, served as a B-24 Liberator navigator in the United States Army Air Forces. He was shot down over Romania and incarcerated in a POW camp, from which he escaped.

- In one scene Stryker instructs bumbling Private Choynski (Hal Baylor) on the correct way to use a bayonet. Ironically, in real life Baylor was an Marine who had fought in the battles of Saipan and Tinian, while Wayne never served in the military.

- The three surviving Iwo Jima flag raisers made a cameo appearance during that scene in the film. The three men made famous by the Joe Rosenthal photograph and who survived the battle for Iwo Jima, Rene Gagnon, Ira Hayes and John Bradley, appear in that scene only. They are seen with John Wayne as he instructs them to hoist the flag (Wayne gives the folded flag to Gagnon).

- Following the success of the movie John Wayne was invited to place his footprints in cement outside Grauman's Chinese Theater, and as part of the event actual black sand from Iwo Jima was flown to Hollywood and mixed into the cement in which the "Duke" left his footprints and fist print.

Reel Marines

- Kirk Douglas was originally cast as Sergeant Stryker.

- John Wayne almost turned the film down, since at forty-two he was rather old for the part, and because he felt the American public had grown tired of war films.

- The movie's credits contain the following historical note: "The first American flag was raised on Mount Suribachi by the late Sergeant Ernest I. Thomas, Jr., U.S.M.C. on the morning of February 23, 1945."

- Two thousand Marines appeared as extras in this movie, according to an article in the *Los Angeles Daily News*.

- The *New York Times* reported that location filming for the movie was shot at Camp Del March, Marine Air Station El Toro, and Marine Corps Base Camp Pendleton, all in Southern California.

- Many of the battle scenes in this movie were taken from "actual combat footage taken at the actual fighting on Tarawa and Iwo Jima" according to a review in *Variety*.

- A made-for-television documentary about the making of this movie was filmed forty-four years after it was made. Entitled *The Making of 'Sands of Iwo Jima'*, it featured interviews with the living cast members.

- The special effects work on this movie included lampblack and oil-covered sand designed to look like the volcanic ash of Iwo Jima.

- John Wayne received his first ever Academy Award nomination for his role as Stryker, and wouldn't be nominated again until he won an Oscar twenty years later for playing Rooster Cogburn in *True Grit*.

Reel Marines

Release date: 23 March 2007
Running time: 124 minutes
Historical context: Peacetime

Tagline: Yesterday was about honor. Today is about justice.

Cast

Mark Wahlberg - Gunnery Sergeant Bob Lee Swagger
Lane Garrison - Lance Corporal Donnie Fenn
Michael Peña - Special Agent Nick Memphis
Danny Glover - Colonel Isaac Johnson
Kate Mara - Sarah Fenn

Quote: "Sir... I don't, I don't feel embarrassed. A Force Recon Marine Scout Sniper disarmed me three weeks out of the academy. If anything... I feel lucky to be alive." - Nick Memphis

Highlights

While the political undertones of Shooter are a bit troubling, there no denying it is an enjoyable movie. The action scenes – in particular those involving the science of sniping – are well executed, and Mark Wahlberg is quite believable as a Marine who knows how to hold 'em and squeeze 'em. Politics aside, at the end of the day Gunny Swagger lives up to his name!

Reel Marines

Plot

Marine Gunnery Sergeant Bob Lee Swagger, a retired Scout Sniper, is one of the few men in the world whose sharpshooting abilities allow him to "take out a target from a mile away." He reluctantly leaves a self-imposed exile in his isolated mountain home at the request of Colonel Isaac Johnson, who appeals to his expertise and patriotism and asks for help in tracking down an assassin who plans on shooting the President from a great distance with a high powered rifle. Johnson gives him a list of three cities where the President is scheduled to visit so Swagger can determine if an attempt could be made at any of them.

Swagger assesses each of the locations and determines that a site in Philadelphia would be most conducive to a long range assassination attempt. He passes this information to Johnson, who purportedly arranges for a response, but this turns out to be a set-up. While Swagger is working with Johnson's agents, including a local police officer, to find the rumored assassin, the Ethiopian archbishop is instead shot while standing next to the President. Swagger is shot by the officer but manages to escape, and the agents tell the police and public Swagger is the shooter and stage a massive manhunt for the injured sniper. Swagger has a stroke of luck, however, when he meets rookie FBI special agent Nick Memphis, disarms him, and steals his car.

Swagger uses the first aid supplies in the car to treat his wounds and escapes by driving into the Delaware River while being chased. He then takes refuge with Sarah Fenn, the widow of Swagger's late spotter and close friend Donnie Fenn, who had been killed years before in a mission in Africa where Swagger himself barely survived. She saves his life by cleaning and stitching Swagger's gunshot wounds, and he later convinces her to help him contact Memphis with information on the conspiracy. Memphis is blamed for allowing Swagger's escape and disciplined for negligence, and independently learns Swagger may have been framed for the assassination by finding several inconsistencies in the evidence and witness statements provided to the FBI by an unnamed federal agency.

When the rogue agents realize their secret has been compromised they kidnap Memphis and attempt to stage his suicide. Swagger tails the agents, and kills Memphis' captors with a scoped .22 rifle equipped with a homemade silencer. He and Memphis then join forces and visit a firearms expert (Levon Helm) in Athens, Tennessee. Together they plot to capture who they think is the real assassin, an ex-sniper allied with Colonel Johnson, but once they find him he commits suicide after revealing the archbishop was the real target and was murdered in order to prevent him from speaking out against U.S. involvement in the genocide of an Ethiopian village. The killings were carried out on behalf of a consortium of American corporate oil interests headed by corrupt Senator Charles Meachum, and Swagger learns that the mission where Fenn was killed was also a part of the operation as they were tasked to cover the withdrawal of the contractors assigned to the job. Swagger records the ex–sniper's confession of his involvement in the African genocide, and then with Memphis' assistance escapes from an ambush by killing twenty-four mercenaries.

Meanwhile other rogue mercenaries have kidnapped Sarah in order to lure Swagger out of hiding. With his new evidence and cat-and-mouse strategy, Swagger and Memphis are able to rescue her when Colonel Johnson and Senator Meachum arrange a meeting to exchange their hostage for the evidence of their wrongdoing. After killing several

Reel Marines

enemy snipers in an isolated mountain range and rescuing Sarah, Swagger and Memphis finally surrender to the FBI.

Later they appear in a closed meeting with the Director of the FBI and United States Attorney General, and Swagger clears his name by loading a rifle round (supplied by Memphis) into his rifle (which is there as evidence, since it was supposedly used in the killing). He aims it at the Colonel and pulls the trigger - and the weapon fails to fire. Swagger explains that every time he leaves his house he removes all of the firing pins and replaces them with slightly shorter ones, thus rendering them unable to fire until he returns.

Although Swagger is exonerated, Colonel Johnson takes advantage of a legal loophole - the Ethiopian genocide is outside American legal jurisdiction - and walks free. The attorney general approaches Swagger and states that as a law enforcement official he must abide by the law, but insinuates that if it was the "wild west" it would be appropriate to clean the system with a gun.

Sometime later the Colonel and Senator plan their next move while at the Senator's vacation house - only to be interrupted by an attack by Swagger. He kills both conspirators, one of the Colonel's aides and two bodyguards, and breaks open a gas valve before leaving. The fire in the fireplace ignites the gas, blowing up the house, and the final scene shows Swagger getting into a car with Sarah and driving away.

Trivia

- ❖ *Shooter* is based on the novel *Point of Impact* by Stephen Hunter.

- ❖ Mark Wahlberg had to lose twenty pounds to give Bob Lee Swagger the slim and ripped look of a field sniper.

Reel Marines

- In the scene after Sarah Fenn first talks to Nick Memphis, Mark Wahlberg is seen standing outside with a Philadelphia Eagles hat and jacket on. Wahlberg played for the Eagles in the 2006 movie *Invincible*.

- Athens, Tennessee, the hometown of the firearms expert Swagger and Nick visited, was the location of the "Battle of Athens," where in 1946 armed citizens removed the corrupt local government and restored free elections.

- Shipped to some theaters under the name *Stars*, this was the final film to play at the landmark Mann National Theater in Westwood, California before it closed on April 20, 2007.

- Swagger's wristwatch is a Suunto Vector, a digital watch made in Finland. In addition to telling time, it has an altimeter, barometer, and digital compass.

- Keanu Reeves was the original choice to play Bob Lee Swagger. According to the movie's script doctor William Goldman Clint Eastwood, Robert Redford and Harrison Ford also passed on the role. These men would have fit the literary Bob Lee Swagger's age a bit more closely than Mark Wahlberg, who was born in 1971. In the book Stephen Hunter had introduced Swagger as a Vietnam veteran, so to accommodate Wahlberg's age this film shows Swagger serving in Africa in the 1990's instead of Vietnam in the 1970's.

- Mark Wahlberg's character's last name of Swagger is derived from the Marine scout/sniper slang term for a quick calculation of a bullet's trajectory, which is SWAG (sophisticated wild ass guess).

Reel Marines

- *Shooter* depicts a number of sniper tactics, thanks to the guidance of former Marine scout sniper Patrick Garrity, who trained Mark Wahlberg for the film. Garrity taught him to shoot both left and right-handed (the actor is left-handed), as he had to switch shooting posture throughout the movie due to Swagger's sustained injuries. He was also trained to adjust the scope, judge effects of wind, do rapid bolt manipulation, and develop special breathing skills. His training also included extreme distance shooting (up to 1,100 yards), and the use of camouflage ghillie suits.

- In real life the shot fired in the assassination would not have hit the archbishop straight on, as happened in the film. When a round is fired it will fall from 30–40 feet depending on the distance of the shot. To compensate, the round is fired in an arch calibrated by how far it is going to fall, the distance of the shot, temperature, humidity, wind and the curvature of the earth. In reality, the shot taken at the archbishop would have come down almost straight on top of his head. In his interview Garrity said, "At 1,800 yards, due to the hydrostatic shock that follows a large caliber, high velocity round such as the 50. Cal, the target would literally be peeled apart and limbs would be flying two hundred meters away." The exit wound on the archbishop's head would have been too graphic for theatres, so instead the movie depicts a more suitable representation of the impact.

- The website that Memphis is given a link to in the chat room, "precisionremotes.com," is real. It is the website of Precision Remotes, a California-based company which designs and manufactures remote-operated weapon and surveillance platforms such as the one in the film.

Reel Marines

- During the mountaintop confrontation Swagger kills one of the snipers by shooting the counter-sniper through his rifle scope. This is likely based upon an infamous kill by renowned sniper Marine Corps Gunnery Sergeant Carlos Hathcock during the Vietnam War.

- The high caliber rifle that Swagger owns and is framed with is a Cheyenne Tactical M200 Intervention, which fires a .408 caliber projectile accurately out to and beyond 2000 meters. The CheyTac M200 is also available with a Long Range Rifle System which consists of a laser range finder, magnifying scope with night vision capability, and a weather-sensing module, all of which interface with a PDA running ballistics calculation software.

- Throughout the film Swagger uses an array of sniper weapons, among which are the USMC M40A3 and Barrett M82 sniper rifles. Donnie Fenn used an M4 Carbine with a M203 grenade launcher and M68 Close Combat Optic in the African opening sequences.

- Some film critics, both liberal and conservative, saw the film as left-leaning in its politics, arguing that the main villain was a clear analogy for Dick Cheney. In fact, the portraits on the walls in the room where the conspirators meet in Langley are all of Republican presidents. In addition, the left-wing website zmag.org can be seen on Swagger's laptop while he is reading the 9/11 Commission report. Zmag publishes articles by, amongst other academics, socialist/anarchist Noam Chomsky.

Reel Marines

SNIPER

Release date: 29 January 1993
Running time: 98 min
Historical context: Peacetime

Tagline: One Shot. One Kill. No Exceptions.

Cast

Tom Berenger - Master Gunnery Sergeant Thomas Beckett
Billy Zane - Richard Miller
J.T. Walsh - Chester Van Damme

Quote: "Let me tell ya somethin' – sittin' in an office giving men orders to kill is the same thing as puttin' a bullet in someone's heart yourself. The same. Goddamn. THING."
- Master Gunnery Sergeant Thomas Beckett

Highlights

Snipers are a different breed, and this film does a good job explaining to the uninitiated what makes them tick. While an artillery barrage or an airstrike can wreak untold destruction, a sniper and his weapon is a precision tool, and with one well aimed shot at the right target he can change the course of a war. Their craft is about much more than shooting however, and that is one of the main points of contention between Beckett and Miller – so pay close attention to the difference in their field craft.

Reel Marines

Plot

The movie begins with Marine sniper Master Gunnery Sergeant Thomas Beckett and his spotter Corporal Papich (Aden Young) assassinating a Panamanian rebel leader in the jungle. Afterward they take up a position in an isolated area and wait until nighttime for an extraction, however they are picked up at daybreak much to the dismay of Beckett. An experienced rebel sniper once trained by Beckett kills Papich as they reach the helicopter, and an infuriated Beckett blames the helicopter pilot for Papich's death.

Beckett is then paired with the very inexperienced, and civilian, Richard Miller. Their mission is to eliminate a rebel general who is being financed by a reclusive Panamanian drug lord. Miller is a SWAT team sharpshooter, but lacks combat experience and has no confirmed kills to his name. When he is on the way to the staging area his helicopter is attacked by a guerrilla with an assault rifle who kills most of the men in it. Miller takes out his rifle and acquires a bead on the man, but is unable to pull the trigger – but fortunately the chopper's dying machine gunner manages to kill the guerrilla before succumbing to his wounds.

When Beckett and Miller step off friction immediately begins between them, based in part on Beckett's insistence on deviating from the mission plan Miller was given, but mostly due to the fact that Miller - who is nominally in command - has no experience in or aptitude for jungle operations. Early on they encounter a group of Indians who agree to lead them past the rebel guerrillas in return for a favor. They want the team to eliminate El Cirujano ("the Surgeon"), a torture master who is working with the rebels and is listed as a target of opportunity.

Reel Marines

Beckett agrees to do so and tells Miller he expects him to make the kill. The general is setting up a meeting with the drug lord, which will allow the sniper team the opportunity to kill both men. Beckett is not certain of Miller's reliability, and he proves him right by deliberately missing the shot. One of the Indians is killed by rebel return fire, and they refuse to help the team any further.

The two men continue to their target and along the way try to locate a village priest who can provide details about the mission, however he has already been tortured and murdered by Alvarez's men before the they can arrive. That night Beckett and Miller set up camp in the jungle and wait until dawn to move out, and as Miller falls asleep Beckett remains on watch. That night the two are tracked by the very sniper who had earlier killed Papich, and Beckett uses Miller as bait to draw the sniper out and kills him before Miller can be harmed. Miller is infuriated.

The two men reach the general's hacienda and take up positions while wearing ghillie suits. While they are waiting for their targets to emerge Miller is seen, and Beckett kills Miller's attacker while Miller shoots and kills the drug lord. The two rendezvous outside the compound, where Beckett insists on going back to kill the general. Miller refuses and an argument erupts, leading to an exchange of fire between Beckett and Miller as the latter follows his orders to eliminate Beckett if he threatens the mission. The shooting ends when Miller runs out of ammunition, and he calms down afterward. As rebels close in Beckett attempts to provide cover fire for Miller and is subsequently captured by rebels, while Miller is able to escape.

That night Miller heads to the base camp where Beckett is being held. Inside he encounters the general and kills him with a knife, and then sees Beckett being tortured and having

Reel Marines

his trigger finger amputated by El Cirujano. Beckett mouths to Miller to shoot both Cirujano and himself with one shot, but instead Miller kills Cirujano and rescues Beckett. The pair then reach their pick-up point and are safely extracted after Beckett saves Miller's life by killing an ambusher.

Trivia

❖ The weapon used by Billy Zane in this movie is not an H&K PSG-1, the expensive precision semi-automatic rifle commonly used by special forces snipers, but instead an H&K SR-9 TC, a similar but cheaper version of the PSG-1 that was designed for civilian sales.

❖ If you look closely at the locomotive of the train which inserts Beckett and Miller, the "QR" symbol of the Queensland Rail can be seen on the side. Queensland is the state in Australia where this movie was shot.

❖ Historically, Thomas Becket was the archbishop of Canterbury in England from 1162 until he was martyred in 1170 during the reign of King Henry II.

❖ This movie, along with several others, uses a story from the life of Carlos Hathcock, who is said to have shot an enemy sniper through his scope and eye.

❖ Beckett is a Master Gunny Sergeant in the Marine Corps, however the crew chief of the helicopter which extracts him calls him "Gunny." A Master Gunnery Sergeant would *never* be addressed as such. The closest appropriate term would be "Master Guns."

❖ *Sniper* spawned two sequels, the TV movie *Sniper 2* in 2002 and the direct-to-video *Sniper 3* in 2004.

Reel Marines

SPACE

Release date: 24 September 1995
Running time: 60 minutes (pilot) / 30 minutes (22 episodes)
Historical context: Future

Tagline: In space, no one can hear you scream, unless it is the war cry of a U.S. Marine!

Cast

Morgan Weisser - Lieutenant Nathan West
Kristen Cloke - Lieutenant Shane Vansen
Rodney Rowland - Lieutenant Cooper Hawkes
Joel de la Fuente - Lieutenant Paul Wang
James Morrison - Lieutenant Colonel T.C. McQueen
R. Lee Ermey - Sergeant Major Frank Bougus

Quote: "Until this war is over, the only easy day is yesterday!"
    - Sergeant Major Bougus

Highlights

This is the second science fiction movie on the list, and it is here for three good reasons. Number one, it features R. Lee Ermey as a Drill Instructor, and you can never go wrong there. Second, the television series it spawned turned out to be pretty good. Finally the third, and most important, rationale for its inclusion is the Leatherneck compiling this list just happens to have been the film's military technical advisor - and I think it was a pretty darn good movie!

Reel Marines

Plot

In the years leading up to 2063 humanity has begun to colonize other planets. Without warning a previously unknown alien species, the "Chigs," attack and destroy Earth's first extra-solar colony and then destroy a second colony ship. The bulk of the Earth's military forces are sent to confront the Chigs but are destroyed or outflanked, and in desperation unproven and under-trained outfits like the 58th Squadron "Wildcards" of the United States Marine Corps Space Aviator Cavalry are thrown into the fight. They are based on the space carrier *USS Saratoga*, and act both as infantry and as pilots of SA-43 Endo/Exo-Atmospheric "Hammerhead" Attack Jet fighters. The Wildcards are the central focus of the movie and subsequent series, which follows them as they grow from untried cadets – trained by Sergeant Major Bouguss, played by none other than R. Lee Ermey - into veterans. Although the unified Earth forces come under the control of a reformed United Nations, the UN has no armed forces of its own and therefore navies such as the U.S. Navy and the Royal Navy operate interstellar starships.

The Space: Above and Beyond milieu includes an underclass race of genetically engineered and artificially gestated humans who are born at the physical age of eighteen, and are collectively known as In Vitroes or, sometimes derogatorily, "tanks" or "nipple-necks." The In Vitroes have replaced the previous underclass, the artificial intelligences known as Silicates. These human-looking androids, referred to as "walking personal computers," have rebelled, formed their own societies, and wage a guerrilla war against human society. The Silicates are also suspected of having some involvement with the Chigs.

Reel Marines

Trivia

- *Space* was the pilot movie for *Space: Above and Beyond*, a short-lived mid-90s science fiction television show on the FOX Network which was created and written by Glen Morgan and James Wong. Originally planned for five seasons, it ran only for the single 1995–1996 season. It was nominated for two Emmys and one Saturn Award.

- According to the producers, the main fictional work which influenced *Space: Above and Beyond* was the 1974 science fiction novel *The Forever War* by Joe Haldeman, in addition to other fictional works such as the 1948 World War II biographical novel *The Naked and the Dead* by Norman Mailer, the 1895 American Civil War novel *The Red Badge of Courage* by Stephen Crane, the *Iliad*, and the 1962 television series *Combat!*.

- *Space: Above and Beyond* shares conspiracy elements with other television shows co-produced by the same team, such as *The X-Files* and *Millennium*.

- One-time characters included Coolio as the Host, an uncredited David Duchovny as Alvin El 1543 (aka "Handsome Alvin"), and retired Marine Captain Dale Dye as Major Jack Colquitt.

- *Space* was filmed in Australia at the Warner Studios on the Gold Coast and at RAAF Base Williamtown.

- Full-scale models of the "Hammerhead" fighters used in the series were being stored on board a freighter before shipping, and crewmen from a Russian freighter were caught taking pictures of them after mistakenly thinking they were a new kind of advanced U.S. tactical fighter.

Reel Marines

- Writer Glen Morgan has two children with actress Kristen Cloke, who plays Lieutenant Shane Vansen. Winslow and Greer Autumn Morgan are named in honor of his *Space: Above and Beyond* characters.

- Marine Master Sergeant Andy Bufalo was the pilot's military technical advisor and had an uncredited part as one of R. Lee Ermey's assistant drill instructors. At the time of filming he was commander of the Marine Security Guard Detachment at the American Embassy in Canberra, Australia.

- The haircuts of some of the male actors were way, *way* outside regulations, but the producers explained to the real Marines that they had a "no haircut clause" in their contracts.

- During the funeral scene the two individuals in Dress Blues folding the flag are not actors, but instead Marines from the Embassy in Canberra. Sergeants Eric Merkel and Joe Hendrix were big fans of R. Lee Ermey, and jumped at the chance to appear in a movie with him.

Reel Marines

Release date: 18 December 2009
Running time: 77 minutes
Historical context: Operation Iraqi Freedom

Tagline: When one falls, another brings him home.

Cast

Kevin Bacon - Lieutenant Colonel Mike Strobl
Tom Aldredge - Charlie Fitts
Blanche Baker - Chris Phelps
Gordon Clapp - Tom Garrett
Mike Colter - Master Gunnery Sergeant Demetry
Ann Dowd - Gretchen Mack

Quote: "If I'm not over there, what am I? Those guys, guys like Chance... they're Marines." - Lieutenant Colonel Mike Strobl

Highlights

The thing which makes Taking Chance different from all of the other "Iraq movies" is it is all realism, and no cynicism. The film is both troubling and inspiring in the spirit of Saving Private Ryan, but on an artfully intimate scale - masterful in technique, but with no self-reverential showing off. Rendering honors is one of the film's themes, and also one of its singular accomplishments. Taking Chance gives us all the opportunity to render honors to the fallen.

Plot

In April of 2004, as casualties mount in Iraq, choices at Quantico focus on either increasing troop strength or only replacing casualties. Lieutenant Colonel Mike Strobl crunches the numbers and, stung by his superior's rejection of his recommendation because he lacks recent combat experience, volunteers for escort duty. He accompanies the remains of nineteen-year-old PFC Chance Phelps from Dover AFB to Philadelphia by hearse, from there to Minneapolis and on to Billings, Montana by plane, and then by car to Phelps' Wyoming home – as person after person pays respects. Kind words, small gifts, and gratitude are given to Strobl along the way for him to deliver to the family on this soul-searching journey.

Perhaps more ceremony than cinema, *Taking Chance* breaks many a movie rule. It barely has a plot, contains the absolute minimum amount of dialogue and lacks the usual antagonist-protagonist conflict. It could also be said that its real hero never speaks, and barely has any scenes.

A road movie in a novel and disquieting sense, *Taking Chance* is "based on actual events," as the opening disclaimer notes. First-time director Ross Katz relies on audio imagery to start the film - sounds of battle that include explosions and cockpit chatter, a blank-screen intercut with quick shots of Kevin Bacon in a military uniform, and of preparations for a long trip to Wyoming and home.

Why are bags of ice carried by workers as if they were precious or sacred artifacts? They are used to help preserve the bodies of slain comrades, packed in temporary flag-draped coffins for a journey that starts at an Air Force base in Germany. In a gesture we are told is unusual, a Marine

Reel Marines

officer will accompany the body of a PFC to its final destination, an assignment for which he volunteers.

His reasons are, superficially, that he and the young Marine are from the same hometown of Clifton, Colorado, although the funeral and burial will be in Dubois, Wyoming due to the divorce of Phelps' parents. It turns out that the veteran and the young Marine share many bonds, and also share the special, profound relationship of men who've been engaged in battle. That battle, during the current war in Iraq, is dealt with here in mainly noncontroversial terms. A photo of George W. Bush dominates a *USA Today* front page at an airport, and a longhaired lad serving as a limo driver says, "I don't really get what we're doing over there" - and that's about it.

Strobl, a seventeen-year veteran of the Marines, feels quietly guilty that he's confined to a desk job and not involved in combat, despite the chestful of medals and decorations which attest to previous combat experience. When he sees the name of Phelps on the latest list of casualties, he is compelled to volunteer for duty as his escort.

He says goodbye to wife and children and embarks on the solemn journey - and that's pretty much the film. We are shown in striking detail many of the little rituals that are part of the larger picture, from the gentle cleansing of the dead man's fingers and toes to the fastening of a barcode tag to the black body bag.

The film makes it clear that remarkable care is shown and taken by those who come in contact with the remains at each stage of the journey, from loading the long box onto one plane or another, carrying it in a hearse, carefully placing a memento inside the coffin and even, in an unusual gesture, Strobl declining the offer of a hotel room so he can sleep near the coffin inside a cargo hangar at a stopover.

Reel Marines

At each stop, whenever the box is transferred from one conveyance to another, Strobl salutes - a smart but slow salute, held until the box passes by. Sometimes onlookers or passersby feel compelled to register their own respect in some similar sort of gesture, and at one point, as the remains are being transported in a long black limousine down a narrow Western highway, the cars behind form an impromptu cortege with their headlights on as another sign of respect.

In the course of Strobl and Phelps' long journey only one citizen shows disrespect - a snippy security agent at an airport checkpoint. Strobl is told to remove his uniform blouse with its heavy load of metal decorations. He angrily refuses, and there is a nonconfrontational resolution after some foolish huffing and puffing by the agent.

Some critics have said the case of young Phelps risks seeming too generic, as if this were a military propaganda piece exploiting a combat casualty, but such reactions are fleeting and finally overwhelmed by the film's heartfelt intensity. The wordplay of the title is perhaps unfortunate, if not seemingly inevitable, and Kevin Bacon's performance is one of his best - stoicism raised to the level of art, yet not without warm, human moments.

"He'd be so happy that an officer brought him home," one of Phelps' relatives tells Strobl at a memorial service, and when Lieutenant Colonel Strobl briefly expresses regret that his role of "escort" feels hollow and meaningless, a Korean War vet reprimands him.

"Without a witness," he tells Strobl, "Chance Phelps would just disappear."

Reel Marines

Trivia

❖ *Taking Chance* is based on the personal journal of Lieutenant Colonel Michael R. Strobl.

❖ The Defense Department banned virtually all media coverage of deceased vets returning home beginning with the 1991 Gulf War through April of 2009, but offered advice and assistance while providing *Taking Chance's* film crew with a rarely viewed but painstakingly accurate account of the care and protocol bestowed upon the nation's fallen warriors.

❖ Lieutenant Colonel Strobl, a Desert Storm veteran, said he decided against another combat tour largely because of his young family - but he was conflicted, and joined the many military personnel who volunteered for escort duty as Iraqi war deaths escalated.

❖ Strobl shared the twenty-page journal of his trip with friends and co-workers via email, and it eventually spread virally to military blogs and the media.

❖ Kevin Bacon, who portrays Strobl, also played a Marine officer in *A Few Good Men*.

❖ The "Marine Driver" in the film was portrayed by real-life Marine Gunnery Sergeant Henry Coy.

❖ Joel de la Fuente, who appears in *Taking Chance* as a ticketing agent, played Marine "space aviator" Lieutenant Paul Wang in *Space: Above and Beyond*.

❖ Tom Wopat, best known for the *Dukes of Hazzard*, appears in the role of Phelps' father.

Reel Marines

TELL IT TO THE MARINES

Release date: 29 January 1927
Running time: 103 minutes
Historical context: China

Tagline: Lon Chaney as a Leatherneck!

Cast

Lon Chaney - Sergeant O'Hara
William Haines - Private George Robert "Skeet" Burns
Eleanor Boardman - Nurse Norma Dale
Eddie Gribbon - Corporal Madden

Quote: "Don't shoot him before we find out where his parents live." - Sergeant O'Hara

Highlights

This was a groundbreaking motion picture in a number of ways. Tell it to the Marines was the prototype for many films about the armed services, in peace and war. A callow youth discovers his backbone in a rigidly structured environment, mentored by an uncompromising father figure. There's a girl in there somewhere, for a futile stab at feminine interest, but the triangle is not about two men and a woman, but about two men and the Marine Corps.

Reel Marines

Plot

"Skeet" Burns, a young slacker with a taste for the race track, joins the Marines simply to get a free train ride to San Diego so he can hop over to Tijuana. He confides his plan to a Marine General traveling in civvies, and the officer tips off Sergeant O'Hara, the tough drill instructor at the base. O'Hara tries to nab Skeet, and loses him, but a few days later a tired, broke Skeet shows up and reluctantly reports for duty. O'Hara puts him through the ringer in basic training, determined to make a man of him.

Skeet takes a fancy to Norma Dale, a Navy nurse at the base, and when he goes off on a tour of duty to a Philippine village she tells him that she just might wait for him. On the boat ride over Skeet picks a fight with a sailor, unaware that he is a boxing champ, and the two put on an exhibition fight for the entire crew - but the experienced boxer makes short work of him.

Once on the island Skeet flirts with Zaya, a native girl, but when he rebuffs her advances her native friends attack him and O'Hara has to dive into the fight to save Skeet's skin. Norma and Skeet later meet in Shanghai, but she has heard about the affair of the native girl and brushes him off. Before they can reconcile, Norma and the other nurses are ordered to Hangchow due to the outbreak of an epidemic. Skeet blames O'Hara for his troubles, and is sent to the brig for coming in after hours one night.

O'Hara releases Skeet when word arrives that the nurses are under siege in Hangchow and every Marine is needed. They are under attack by a band of Chinese bandits, and the Marines must defend a bridge until reinforcements arrive. O'Hara is wounded, but in the end Skeet pulls him through and demonstrates true bravery under fire.

Reel Marines

Trivia

- ❖ The phrase "Tell it to the Marines" suggests that you don't really believe someone. The story goes that King Charles II, being told by a naval officer that such things as flying fish existed, remarked "Tell that to my Marines." A nearby Marine officer, who felt that this was an insult, was mollified when the King explained it was a compliment because his Marines had been to all parts of the world and had seen everything – therefore if they had not seen flying fish, then they didn't exist.

- ❖ Lon Chaney often stated that this was his favorite film. It was the biggest box office success of Chaney's career, and the second biggest moneymaker of 1927.

- ❖ Not only did Lon Chaney forgo his customary grotesque makeup for this picture, he refused to wear any film makeup at all because he felt to have done so would have detracted from the documentary reality and integrity of the picture.

- ❖ MGM brought in General Smedley Butler, commander of the Marine base in San Diego, for technical consultation on the film. Lon Chaney formed a close friendship with the General which lasted for the rest of Chaney's life.

- ❖ The studio was allowed to shoot on the base, which made *Tell It to the Marines* the first motion picture made with the full cooperation of the Marine Corps.

- ❖ Battleship *USS California,* which was later involved in the attack on Pearl Harbor on December 7, 1941, was used for the scenes at sea.

Reel Marines

❖ The final scenes of the film where the Marines rescue the hostages were filmed at Iverson's Ranch in Chatsworth, California, the location for such films as *Fort Apache* and *The Good Earth*.

❖ A writer in *Leatherneck* magazine wrote that "few of us who observed Chaney's portrayal of his role were not carried away to the memory of some sergeant we had known whose behavior matched that of the actor in every minute detail ..."

❖ Warner Oland appears as the Chinese bandit leader at the film's end and was called "a wicked scene-stealer" in a very memorable bit.

❖ William Haines, who was cast as "Skeet," lived an openly gay lifestyle in Hollywood.

❖ For his role in the film, Lon Chaney became an Honorary Marine - the first film star to do so.

Reel Marines

THE BOYS IN COMPANY C

> Release date: 20 April 1978
> Running time: 125 minutes
> Historical context: Vietnam
>
> Tagline: To keep their sanity in an insane war, they had to be crazy.
>
> **Cast**
>
> Stan Shaw - PFC Tyrone Washington
> Andrew Stevens - Private Billy Ray Pike
> James Canning - Private Alvin Foster/Narrator
> Michael Lembeck - Private Vinnie Fazio
> James Whitmore, Jr. - Lieutenant Archer
> R. Lee Ermey - Staff Sergeant Loyce
> Santos Morales - Staff Sergeant Aquilla
>
> Quote: "Goddamnit, oh-three-hundred is basic infantryman. Oh-three-hundred is the United States Marine Corps!"
> - Staff Sergeant Loyce

Highlights

The Boys in Company C sometimes stretches the limits of credulity, but there are just enough highlights, like R. Lee Ermey in his first role (notice a pattern?) and Staff Sergeant Aquilla's ranting (see "Trivia" for a sample) to get it on the list. "Reach down and grab your nuts. I said grab them! Do you like them? Do you want to keep them? Then listen to what we tell you!"

Reel Marines

Plot

The Boys in Company C is a 1978 film about United States Marines during the Vietnam War. This war drama, which prefigures the later *Full Metal Jacket*, follows the lives of five young Marine Corps inductees from their training in boot camp in 1967 through a tour in Vietnam in 1968 - which quickly devolves into a hellish nightmare.

Disheartened by futile combat, appalled by the corruption of their South Vietnamese allies, and constantly endangered by the incompetence of their own company commander, the young Marines find a possible way out of the war. They are told that if they can defeat a rival soccer team, they can spend the rest of their tour playing exhibition games behind the lines - but as they might have predicted, nothing in Vietnam is as simple as it seems.

Trivia

* *The Boys in Company C* was among the first Vietnam War films to appear after the Vietnam Era.

* The part of Staff Sergeant Loyce was the first role for R. Lee Ermey of *Full Metal Jacket* fame.

* The original script was written by Rick Natkin for a film class at Yale University in 1973.

* When SSgt Loyce (Ermey) picks up the platoon of recruits at receiving he is a Sergeant, but later on he is seen as a Staff Sergeant with no explanation as to why or when he was promoted.

* Stan Shaw, who played Tyrone Washington, appeared in *The Great Santini* as Ben Meecham's friend "Toomer."

Reel Marines

- The movie was filmed in the Philippines.

- The dialogue for which this movie is best remembered:

    SSgt Aquilla: You with the pretty face! What the hell are you spitting on the ground for? What is that shit in your mouth?

    Pvt Billy Ray Pike: Chewing tobacco sir.

    SSgt Aquilla: It's *what?*

    Pvt Billy Ray Pike: Chewing tobacco sir!

    SSgt Aquilla: What the hell are you chewing tobacco for!?

    Pvt Billy Ray Pike: I chew when I play baseball sir!

    SSgt Aquilla: Does this look like a baseball stadium to you?

    Pvt Billy Ray Pike: No Sir, it doesn't.

    SSgt Aquilla: Then what the hell are you chewing tobacco for? Did you get on the wrong fucking train, or did they draft you?

    Pvt Billy Ray Pike: I enlisted Sir!

    SSgt. Aquilla: You did *what?*

    Pvt Billy Ray Pike: I enlisted in the Marines, Sir!

    SSgt Aquilla: What the hell would you do a crazy thing like that for?

    Pvt Billy Ray Pike: [pause]

    SSgt Aquilla: Answer me goddamn it!

    Pvt Billy Ray Pike: I want to fight for my country sir!

    SSgt Aquilla: Fucking TURD!

Reel Marines

THE D.I.

Release date: 30 May 1957
Running time: 106 minutes
Historical context: Peacetime

Tagline: Give him 12 weeks. He'll give you a Few Good Men.

Cast

Jack Webb - Tech Sergeant Jim Moore
Don Dubbins - Private Owens
Lin McCarthy - Captain Anderson
Paul E. Prutzman - Sergeant Braver

Quote: "Private Owens! Was the sand flea you killed male or female?" - Technical Sergeant Jim Moore

Highlights

Long before R. Lee Ermey came onto the Hollywood scene Jack Webb was the actor most closely identified with the role of Drill Instructor. One of the things which makes Webb's performance in this film truly remarkable is even though he does not utter a single bit of profanity, he was so convincing rumors circulate to this day which claim him to be the real article. Countless Marine veterans have watched this movie and felt a bit of nostalgia, perhaps because of the old time uniforms, or maybe because it was filmed in black and white - and that is why it is a classic.

Reel Marines

Plot

Technical Sergeant Jim Moore is one of the toughest Drill Instructors on Parris Island, but he's got a thorn in his side – a recruit named Private Owens, who always seems to foul up when the pressure is on.

Moore is convinced that "there's a man underneath that baby powder," and drives Owens to the point of desertion. To make matters worse the Series Officer, Captain Anderson, has given Moore just three days to make the scared private into Marine Corps material, "or I'll personally cut the lace off his panties and ship him out!"

In the most memorable scene in the film the recruits are maneuvered into a hasty ambush while on a route march, and as they lay there in wait for an imaginary enemy Owens slaps at a sand flea which is crawling on his neck. His lack of discipline infuriates Moore, who tells him that his weakness could have given away their position and cost the lives of every man in the platoon if they had been in combat.

The platoon is then tasked with finding the dead sand flea - the exact one which Owens has killed - in order to give it "a proper funeral." While most of the recruits search in the dark on their hands and knees, others dig a full size grave, and there is even a bit of comic relief when one of the hapless privates tried to convince Moore he has found "the" sand flea. Moore asks him if it is male or female, and naturally the Drill Instructor is prepared to tell the young recruit the murdered flea was the *other* sex – no matter what he says.

In an attempt to balance the movie somewhat and show that Moore has a personal life off Parris Island - and adding to the pressure on him - he also juggles a budding romance with a shop girl in a minor subplot.

Reel Marines

Trivia

❖ *The D.I.* was shot in just twenty-three days during March of 1957.

❖ The movie was also produced and directed by star Jack Webb.

❖ To make the summer release date requested by Warner Brothers, Webb edited as he shot.

❖ Only three of the men in this film were professional actors - Webb, Dubbins and Lin McCarthy. All of the others were actual Marines.

❖ Moore's rank has confused many modern-day Marines, some of whom believe he is actually wearing Army chevrons. Technical Sergeant was a rank in the Marine Corps until 1958. From 1941 until 1946 it was equivalent to "grade two," ranking with Gunnery Sergeant and other technical ranks with which it shared its insignia. From 1947 until 1958 the grade was reclassified as E-6, and became the sole rank in this grade. It was renamed Gunnery Sergeant and elevated to E-7 after the reorganization of grades in 1959.

❖ Webb was married to actress and singer Julie London from 1947-54. She later married a real-life Marine, actor and songwriter Bobby Troup, who played a Marine in the 1967 film *First to Fight*.

❖ The film was based on a Kraft Television Theater presentation called *Murder of a Sand Flea*. Lin McCarthy played the same role in both productions.

Reel Marines

THE GREAT SANTINI

Release date: October 26, 1979
Running time: 115 minutes
Historical context: Peacetime/Cuban Missile Crisis

Tagline: The bravest thing he would ever do was let his family love him.

Cast

Robert Duvall - Lieutenant Colonel Wilbur P. "Bull" Meechum
Blythe Danner - Lillian Meechum
Michael O'Keefe - Ben Meechum
Stan Shaw - Toomer Smalls
David Keith - Red Petus
Paul Mantee - Colonel Virgil Hedgepath

Quote: "You're gonna hack it, or pack it!" - "Bull" Meechum

Highlights

If ever an actor has captured the essence of a Marine on film, it is Robert Duvall as Lieutenant Colonel Wilbur P. "Bull" Meecham. Non-Marines will watch this movie and tut-tut about his perceived shortcomings as a human being, but Marines will smile because they recognize the embodiment of the warrior spirit when they see it. Some may call him brash, cocky, arrogant and overbearing, but I say, "Stand by for a fighter pilot!"

Reel Marines

Plot

The *Great Santini* tells the story of a senior Marine Corps officer whose success as a military aviator contrasts with his shortcomings as a husband and father, while exploring the high price of heroism and self-sacrifice. The four Meechum kids and their mother move from Marine post to Marine post following their father, who is fighter pilot in the peaceful years before the Vietnam War. Bull Meechum, the self-described "Great Santini," is manic - a martinet who is at once enthusiastic and abusive of his family. Bull loves fighting almost as much as he loves the Marine Corps. Profane, cocky, and arrogant, he's a great fighter pilot - and he knows it.

When he arrives in Beaufort, South Carolina Bull is assigned to whip a squadron of pilots into shape. His Commanding Officer, who is an old rival, hates his guts, but knows that if he's going to straighten out his lagging squadron Meechum is the man to do it. In the meantime son Ben makes friends with Toomer, a slow-talking local black youth. Racial and family tensions explode into violence, and Ben must find a way to make peace with his father.

The story and irony of *The Great Santini* is in Meechum's total intolerance of family life and fatherhood. He has a lovely, supportive wife in Lillian, an earnest, likeable son in Ben, three smaller children and a good home - but Meechum finds the pastoral nature of peacetime totally incompatible with his gung-ho nature. He begins to drink, and drills his family unmercifully like recruits. He also hammers his son relentlessly until, in a basketball game, his Ben fights back and the family cheers his efforts. Tension builds in the household until, during one drunken night, Meechum breaks down and reveals what really drives him.

Reel Marines

Trivia

- ❖ The script was adapted by Lewis John Carlino from the novel of the same name by Pat Conroy with assistance from an un-credited Herman Raucher.

- ❖ The title character, Lieutenant Colonel Wilbur "Bull" Meechum, was based on Conroy's father Donald Conroy, a Marine fighter pilot who referred to himself in the third person as "The Great Santini."

- ❖ The story, for the most part, follows the book. As with most movies based on books, there is some material left out. The movie's major divergence is the absence of Sammy, Ben Meechum's Jewish best friend. The spelling of the family name is also changed from Meecham to Meechum.

- ❖ Warner Brothers released this film in 1979 for a brief time under an alternate title, *The Ace*.

- ❖ Much of the film was shot on location in Beaufort, South Carolina.

- ❖ The setting of the Meechum house was later used for *The Big Chill*.

- ❖ The character of "Coach Spinks" is based on Jerry Swing, who coached Pat Conroy in high school.

- ❖ Despite playing their mother, Blythe Danner is only twelve years older than Michael O'Keefe and Lisa Jane Persky.

- ❖ Blythe Danner is the real-life mother of actress Gweneth Paltrow.

Reel Marines

- On the morning of the Academy Award nominations in 1981, when the movie got nods for best actor and best supporting actor, author Pat Conroy received a phone call from his father who told him, "You and me got nominated for Academy Awards, and your mother (his ex-wife) didn't get squat."

- In the opening dogfight sequence you hear the lead pilot call "Okay, it's Marines verses Navy," and you see an F-4J Phantom with the word Navy on the side, but the squadron designation is VMFA-251. Only Marine Phantoms carried the "VMFA" designation, with Navy Phantoms being "VF." The "Marine" squadron was in fact the VMFA-251 Thunderbolts, who now fly the F/A-18C Hornet.

- The squadron Bull assumes command of, VMFA-312, were actually known as the "Checkerboards," not "Werewolves" as depicted in the film.

- Stan Shaw (Toomer) played Marine Tyrone Washington in the *The Boys in Company C.*

- The part of redneck Red Pettus was actor David Keith's film debut. Later in his career he would have to contend with a Marine Drill Instructor by the name of Gunny Foley in *An Officer and a Gentleman.*

- Paul Mantee, who played Colonel Hedgepath, refused to get a regulation Marine Corps haircut when approached by the film's technical advisor – in contrast to the squared away high and tight of star Robert Duvall.

Reel Marines

Release date: 13 October 2006
Running time: 92 min
Historical context: Iraq War

Tagline: A one man strike force who never surrenders.

Cast

John Cena - John Triton
Robert Patrick - Rome
Kelly Carlson - Kate Triton
Abigail Bianca - Angela

Quote: "You married a Marine, Kate." - John Triton

Highlights

Yes, the acting is marginal at best, and yes, the plot is somewhat predictable, but this motion picture does have a few things going for it. First of all, how in the world would it be possible to exclude a film with the title "The Marine"? Next, Jon Cena has gone a long way towards convincing people that his physique is typical for a Marine – and more than a few Devil Dogs have been inspired by him to hit the gym a little bit harder. Finally, this is an action film, plain and simple, and what in the heck is the Marine Corps about if not action?

Reel Marines

Plot

Marine John Triton is at war in Iraq when he learns that his buddies have been captured and are about to be killed. Triton's commander is going to send help, but it's going to take too long so Triton decides to kill the terrorists himself. Once Triton has returned to the U.S. he and his wife Kate decide to take a vacation. Meanwhile, criminal mastermind Rome robs a jewelry store with his accomplices - girlfriend Angela, Morgan, Vescera, and Bennett. While running from the cops they stop at a gas station, where John and Kate have stopped as well. When a patrol car arrives to fill up, Morgan murders one of the officers, and Rome wounds a second. Bennett knocks John out with a fire extinguisher, and Kate is kidnapped. Morgan then shoots at the gas tanks, causing the store to blow up with John inside - but he survives, emerging to take the abandoned patrol car. He gives chase to the edge of a lake where, while under heavy fire, John falls out of the car and into the water.

The criminals decide to travel through swamps to avoid the police, with Kate still as their hostage. John revives, and despite a police officer denying him permission, continues his pursuit. After an altercation between Morgan and Vescera which shows how crazy Morgan is, Rome decides they have no further use for Vescera and shoots him, leaving the body for the alligators. The criminals eventually arrive at a shack and rest, and Triton gets delayed when some fugitives attack him. He subdues them both and tracks the gang to the lodge where he kills Morgan, who was sent out to turn on the electricity, with the knife Vescera dropped when he was killed. He then kills Bennett when he checks on Morgan, beating him down and stomping on his neck, and as he drags the bodies under the shack the police officer he met

Reel Marines

earlier "grants" him further permission to do what he must and turns off the electricity as a distraction.

Kate takes advantage of this and rushes out of the building, only to be pursued and captured by Angela. John crashes through the window and meets Rome for the first time, and the cop also enters the room - and points his gun at Triton. He has been working with Rome the whole time, and turns out to be the mystery individual Rome tried to cut out of the deal in an earlier phone call. Rome opens fires on Triton, and uses the corrupt cop as a human shield. Rome makes his escape and joins Angela and the still captured Kate before firing at a gasoline tank and destroying the shack as John makes a narrow escape out the window.

Rome takes the cop's car and drives off, but because of the police monitor they have to abandon it, so Angela hitchhikes and shoots a truck driver to steal his vehicle. John is arrested by an officer in a boat, but steals the vessel after handcuffing him and heads for the destination he heard the criminals speaking of earlier – Rita's Marina. John arrives, sees the truck, and jumps onto the back. He then sees a bus heading in the opposite direction and traverses the running board to the passenger seat where he opens the door, grabs Angela by the neck, and throws her into the bus's windshield - killing her and spilling the diamonds. Rome then knocks John off the truck by driving into the side of a building, careens through a warehouse, and leaps out before the truck crashes through a window and into the lake beyond. Triton confronts Rome in the fiery warehouse and thinks he has killed him by burying him in debris, and rushes to rescue his wife who is still handcuffed inside the sinking truck cab. He drags her from the water and successfully administers CPR, but a badly-burned Rome returns and chokes him with a chain until Triton turns the tables and breaks Rome's neck.

Reel Marines

Trivia

❖ *The Marine* stars professional wrestler John Cena.

❖ The movie was produced by the films division of World Wrestling Entertainment, called WWE Films, and distributed in the United States by 20th Century Fox.

❖ The script for *The Marine* was originally written with Academy Award-winner Al Pacino in mind for the part of Rome, and Stone Cold Steve Austin as Triton. After Austin and the WWE parted ways in 2004 the role of John Triton was given to John Cena, and Pacino turned down the role of Rome due to the low salary offered.

❖ After Pacino turned the role down Patrick Swayze was considered for the part of Rome, and it was eventually given to Robert Patrick.

❖ Principal photography for the film was shot and completed in 2004. In order to cover for John Cena's absence from WWE events a storyline was written after Carlito "beat" Cena for the United States Championship saying he had been stabbed by Jesús, the bodyguard of Carlito, and was taking time off to recover.

❖ In one scene Bennett comments that John Triton is like the Terminator. Rome flashes Bennett a look, as Robert Patrick has once played the T-1000 liquid metal robot which was defeated by the Terminator Model 800.

❖ Primary filming was done at Movieworld on the Gold Coast in Queensland, Australia, and the Marine Base was filmed at Bond University. The part in which skyscrapers appear was shot in the downtown area of Brisbane.

Reel Marines

- ❖ A sequel, *The Marine 2,* was released in December of 2009. Unlike the original, it was released direct-to-DVD. Also unlike the original, it is rated R instead of PG-13.

- ❖ *The Marine 2* is inspired by true events. It was shot entirely on location in Phuket, Thailand and stars WWE star Ted DiBiase as Marine Recon sniper Joe Linwood. Reminiscent of the 2001 Dos Palmas incident, Linwood's trip to paradise with his wife is suddenly shattered when guerillas take control of a secluded five-star resort and ask for a substantial ransom. The guerilla leaders begin murdering hostages as the clock ticks away, forcing Joe to use his expert skills to save the hostages from certain death.

- ❖ WWE star Randy Orton, who is a former Marine, had originally been scheduled to play the lead role in *The Marine 2* but could not due to a collarbone injury.

Reel Marines

THE OUTSIDER

Release date: December 1961
Running time: 108 minutes
Historical context: World War II

Tagline: The story of Ira Hayes, who hit the heights at Iwo Jima!

Cast

Tony Curtis - Ira Hamilton Hayes
James Franciscus - Private James B. Sorenson
Gregory Walcott - Sergeant Kiley
Bruce Bennett - General Bridges

Quote: "It's like bein' in a halloween costume you can't get out of." - Ira Hayes

Highlights

Back in 1961 combat veterans didn't have Post Traumatic Stress Disorder, they were "shell shocked," and while Ira Hayes certainly did suffer from PTSD he also had a serious case of survivor's remorse. Some people have speculated that if he had remained in the Marine Corps, rather than returning to the reservation, he would have benefitted from the sort of camaraderie which can only found in the company of others who have witnessed combat firsthand. What Ira Hayes did will always be remembered, but what he went through should never be forgotten.

Reel Marines

Plot

The Outsider is based on the tragic life of Pima Indian Ira Hamilton Hayes, one of the Marines who famously raised the U.S. flag on Iwo Jima.

During World War II seventeen-year-old Ira Hayes, a shy Pima Indian who has never set foot outside his tribal reservation in Arizona, enlists in the Marine Corps. Although most of his white companions either deride or ignore him, he strikes up a deep and lasting friendship with another Marine named Jim Sorenson.

In February of 1945 the two buddies are among the five Marines who raise the U. S. flag on Mt. Suribachi during the bloody fighting at Iwo Jima. Shortly thereafter Sorenson is killed by enemy fire, and a stunned and heartbroken Ira is returned to the United States to take part in a war bond drive. Disturbed at being singled out as a national hero, and feeling unworthy of the role, the simple Indian turns to whisky for courage. His drinking becomes a scandal, and he is returned to his unit in disgrace.

After the war ends anonymity eludes him as his tribal chief persuades him to go to Washington to seek funds for an irrigation project, and while there he begins drinking again and lands in jail.

The dedication of the Iwo Jima Memorial in Arlington, Virginia inspires him to pull himself together, and he returns home to work on the reservation - but is shattered when his people do not elect him to the tribal council. Then one night after a card game he sneaks away with a bottle of liquor and seeks refuge on a lonely mountainside. There he dies of exposure at the age of thirty-two.

Reel Marines

Trivia

- ❖ Ira was born on the Gila River Indian reservation near Sacaton, Arizona. There is now a statue of him there.

- ❖ Hayes died on the reservation January 24, 1955 and was buried in Arlington National Cemetery.

- ❖ Because he was trained as a Para-Marine, Hayes was dubbed "Chief Falling Cloud."

- ❖ The film is based upon author William Bradford Huie's story *The Outsider*, which was published in 1959 as part of his collection *Wolf Whistle and Other Stories*.

- ❖ In the 1960 telefilm *The American*, the part of Hayes was played by Marine veteran Lee Marvin.

- ❖ Ira Hayes appeared as himself in the 1949 John Wayne film *Sands of Iwo Jima*.

- ❖ Critics generally agreed that Tony Curtis was terribly miscast as Hayes – but he still turned in an excellent performance.

- ❖ *The Ballad of Ira Hayes,* written by folk singer Peter La Farge, was recorded many times - with the most popular version being by Johnny Cash.

- ❖ *The Outsider* was the film debut of Lynda Day, who had a small, uncredited role as "Kim." She would marry Marine veteran Christopher George nine years later and change her name to Lynda Day George.

Reel Marines

THE PACIFIC

Release date: 14 March 2010
Running time: 600 minutes (60 per episode)
Historical context: World War II

Tagline: Because our cause is just.

Cast

Joseph Mazzello - Corporal Eugene "Sledgehammer" Sledge
Jon Seda - Gunnery Sergeant John Basilone
James Badge Dale - PFC Robert Leckie
William Sadler - Colonel Lewis "Chesty" Puller
Ben Esler - PFC Charles "Chuck" Tatum

Quote: "Do you want to live? Get off the beach!" – Gunnery Sergeant John Basilone

Highlights

The story of the war in the Pacific has been told in bits and pieces over the years, one battle at a time, but never before has a sweeping epic been produced which follows the Marine Corps through the course of the war. Icons such as Chesty Puller and Manila John Basilone are featured prominently in The Pacific, but so are a couple of lesser known grunts who, luckily for us, happened to have a way with words. This movie is to the Corps what Saving Private Ryan and Band of Brothers were to the Army, and it should not be missed by anyone.

Reel Marines

Plot

The Pacific is a 2010 ten-part television World War II miniseries which was produced by HBO, Seven Network Australia and DreamWorks. It premiered on March 14, 2010.

The film is based primarily on the memoirs of two U.S. Marines - *With the Old Breed* by Eugene Sledge, and *Helmet for My Pillow* by Robert Leckie. The series tells the stories of the two authors along with that of legendary Marine John Basilone as the war against Japan rages.

On December 8, 1941, just over twenty-four hours after the Japanese surprise attack on Pearl Harbor, Congress issued a formal declaration of war against the Empire of Japan. Practically overnight military recruiting offices across the country were jammed as thousands of Americans rushed to enlist in the armed forces. Many of those young men chose to join the Marine Corps, which saw its ranks more than triple in the six months following Pearl Harbor.

The Pacific depicts the war a world away in the Pacific Theater of Operations, which encompassed most of the Pacific Ocean and its islands, including the Philippines, the Netherlands East Indies, New Guinea and the Solomon Islands. The script follows the intersecting odysseys of three men of the 1st Marine Division, an infantry division nicknamed "The Old Breed" for its position as the oldest and largest active duty division of the U.S. Marine Corps, which was at the forefront of many of hardest-fought campaigns of the Pacific War.

Private First Class Robert Leckie grew up in Rutherford, New Jersey as one of eight children, and began a professional sportswriting career for the *Bergen Evening Record* newspaper at age sixteen. He would be christened "Lucky" by his comrades in arms, and was one of those who

Reel Marines

enlisted in the Marine Corps just after Pearl Harbor. He served with H Company, 2nd Battalion, 1st Marine Regiment, 1st Marine Division as a machinegunner.

Sergeant John Basilone was raised in Raritan, New Jersey as one of ten children of Italian immigrant parents. In 1934, at age eighteen, Basilone enlisted in the U.S. Army and served three years in the Philippines, where he was a champion boxer. After a brief return to New Jersey Basilone enlisted in the Marine Corps in 1940 and served as a machinegunner with C Company, 1st Battalion, 7th Marine Regiment, 1st Marine Division, and later with the B Company, 1st Battalion, 27th Marine Regiment, 5th Marine Division.

Born to a privileged family in Mobile, Alabama, PFC Eugene Sledge had relatives on both sides of his family who had fought for the Confederacy. Sledge, the son of a physician who was a medical officer during the First World War, had turned eighteen just one month before the U.S. entered the war, but a heart condition kept him from enlisting until December of 1942. Although his family urged him to train as an officer, Sledge ultimately joined as an enlisted man and served with K Company, 3rd Battalion, 5th Marine Regiment, 1st Marine Division as a mortarman.

Over the span of ten hours *The Pacific* takes an unflinching "under the helmet" look at the experiences of these men and their brothers in arms, each of whom finds himself fighting for his life on faraway specks of land they had never heard of like Guadalcanal, Cape Gloucester, Peleliu, Iwo Jima, and Okinawa. Forced to endure extreme deprivation and a debilitating climate, while fighting a brutal enemy who would rather die than consider surrender, these Marines are driven to the brink of their humanity.

Reel Marines

Trivia

❖ *The Pacific* is based primarily on the memoirs of two Marines - *With the Old Breed* by Eugene Sledge, and *Helmet for My Pillow* by Robert Leckie.

❖ The executive producers for the project were Tom Hanks, Steven Spielberg and Gary Goetzman.

❖ *The Pacific* was filmed as a miniseries for HBO, cost $200-million dollars to make, is ten hours in length, and required seven years of planning and production.

❖ To prepare for filming, screenwriter Bruce C. McKenna accompanied a locations crew to the tiny coral island of Peleliu.

❖ A ridge on Peleliu is laced with hundreds of caves, undisturbed for more than half a century, where Japanese troops hid out from U.S. Marines during one of the war's deadliest conflicts. "There are still skeletons in the caves, and we saw them," McKenna remembered. "At the first cave we found, we walked in and there was the rib cage of a dead Japanese soldier. Up in the hills, every square inch is covered with shell casings and rusted machine guns. The place is unbelievable."

❖ Unlike Iwo Jima, Okinawa and Guadalcanal, whose names still ring in the popular lexicon, Peleliu is largely unremembered, a fact troubling to surviving veterans who fought there.

❖ A full quarter of the series, two and a half of the ten hours, unfolds on Peleliu, with less than a single hour on Iwo Jima.

Reel Marines

- Although part of the film takes place during the battle of Iwo Jima, there is no depiction of the famous raising of the flag atop Mt. Suribachi.

- Because Clint Eastwood had already well acquitted Iwo Jima in *Flags of Our Fathers* and *Letters From Iwo Jima*, Iwo did not become the centerpiece of the series.

- Producer Steven Spielberg said, "When I told my dad we were doing this, he only said one thing… 'Atta-boy!'"

- Much of the film was shot in Melbourne, Australia near where U.S. Marines were stationed during the war.

Reel Marines

THE ROCK

Release date: June 7, 1996
Running time: 136 minutes
Historical context: Peacetime

Tagline: Alcatraz. Only one man has ever broken out. Now five million lives depend on two men breaking in.

Cast

Sean Connery - John Patrick Mason
Nicolas Cage - FBI Special Agent Stanley Goodspeed
Ed Harris - Brigadier General Francis X. Hummel
Tony Todd - Captain Darrow
John Spencer - FBI Director James Womack
David Morse - Major Tom Baxter
Michael Biehn - Commander Anderson

Quote: "Me and my boys are cocked, locked and ready to rock." - Captain Frye

Highlights

In *The Rock* the "bad guys" are the Marines, but while their tactics are questionable at best, their motives are certainly pure. It's never a good thing when brother goes against brother, but most Marines will still have a hard time not smiling with smug satisfaction when the Navy SEAL team led by Michael Biehn is annihilated by a rogue group of Force Recon operators.

Reel Marines

Plot

A group of rogue Force Recon Marines led by heroic, but disenchanted, Brigadier General Francis X. Hummel steals a stockpile of rockets armed with VX nerve agent. They then seize Alcatraz Island during a guided tour, take eighty-one tourists hostage, and threaten to launch the rockets against the population of San Francisco unless the government pays reparations, including compensation, to the families of Marines who have died on clandestine missions.

The Pentagon and Federal Bureau of Investigation decide to deploy a Navy SEAL team to retake the island and free the hostages, but need firsthand knowledge of the underground tunnels of Alcatraz - and are forced to release imprisoned former MI6 and British Army SAS operative John Patrick Mason, the only inmate of Alcatraz who has ever successfully escaped, and who has been in prison for the past thirty years without a trial because he was accused of stealing the private files of J. Edgar Hoover. FBI chemical weapons expert Dr. Stanley Goodspeed, a biochemist who works for the FBI, is also recruited to neutralize the VX nerve agent threat, and the two men and the SEAL team commence their raid on Alcatraz while the Pentagon prepares a secondary countermeasure to Hummel's rockets - thermite plasma bombs capable of incinerating the nerve agent, along with anyone left on the island.

The SEAL team successfully infiltrates Alcatraz, but is ambushed by Hummel's Marines as they enter the main facility. After a tense standoff a shootout ensues in which all the SEALs are killed, leaving only Mason and Goodspeed. Goodspeed convinces a disillusioned Mason to help him finish the mission, and the two begin to move through Alcatraz looking for the VX weapons.

Reel Marines

Mason and Goodspeed manage to remove the guidance chips from thirteen of the fifteen missiles, rendering them useless. This continues until finally they are captured close to dawn and left unguarded in holding cells, and Mason frees them both with one hour and two missiles remaining.

Meanwhile the thermite plasma weapons are readied and armed F-18s approach Alcatraz. The deadline passes, and the Pentagon calls Hummel and asks for another hour to transfer the money. In response the Marines fire one of the two remaining rockets at the Oakland Raiders game, but Hummel discreetly redirects the missile and it detonates harmlessly at sea. Having balked at launching against civilian targets, Hummel reveals the threat to be a bluff. His men agree to end the stand-off, but Captain Frye and other mercenary Marines decide to fire the remaining missile. A short gun-battle ensues in which Hummel and his loyalists are shot dead, with Goodspeed and Mason looking on. They drag the dying Hummel away from the fight, and he tells Goodspeed the location of the last rocket.

Goodspeed manages to locate and disarm the last missile, but he is attacked by Frye in the process. Outmatched, Goodspeed forces a VX pearl into Frye's mouth, killing him. He then injects the atropine antidote into his own heart to save himself from the effect of the VX pearl. Weakened, he scrambles to open ground to signal the command center to abort the airstrike. His green flares are spotted at the last moment, but the abort command is received too late by the pilots and one of them drops the first of his thermite plasma weapons onto the island. The bomb misses the hostages' cell block, but Goodspeed is thrown face down into the water. Mason reappears at this moment, pulling the unconscious Goodspeed to shore, and when the FBI Hostage Rescue Team arrives to secure Alcatraz Goodspeed informs them

Reel Marines

that the escaping Mason was actually "vaporized" by the explosion - releasing him to freedom, anonymity, and his estranged daughter. Before his departure, however, Mason tells Goodspeed the location of the secret microfilm he had stolen, and the movie ends with Goodspeed and his pregnant bride Carla recovering the microfilm along with half a century of state secrets.

Trivia

- Sean Connery's character, John Mason, insists on having a suite at the Fairmont Hotel in San Francisco, which is located at 950 Mason Street.

- The scene in which FBI director Womack is thrown off the balcony was filmed on location at the Fairmont, and numerous calls were made to the hotel by people who saw a man dangling from the balcony.

- It was Nicolas Cage's idea that his character would not swear. His euphemisms include "gee whiz."

- Sean Connery insisted the producers build him a cabin on Alcatraz because he didn't want to travel from the mainland to the island every day. He got it.

- Most of the scenes involving F/A-18s are stock footage of the U.S. Navy Blue Angels.

- Stanley Anderson, who is uncredited as the President, also plays the President in *Armageddon*, another Michael Bay/Jerry Bruckheimer film. In both films there is a scene where he stands in silhouette against a bright window while contemplating a decision which could doom the heroes.

Reel Marines

- Cinematographer John Schwartzman is one of Nicolas Cage's cousins. His mother, *Godfather/Rocky* actress Talia Shire, is Cage's aunt.

- Ed Harris, who plays Gereral Hummel, also co-starred with Michael Biehn in *The Abyss*. In that film it was Biehn's character who was a renegade, and Harris' character was attempting to stop him.

- The picture of General Hummel in Vietnam was taken from *Borderline,* one of Harris' early movies.

- Harris plays a fictional Marine General in this film, but in *The Right Stuff* he portrayed a real Marine Corps Colonel named John Glenn.

- At the beginning of the movie General Hummel kisses his wife's gravestone and leaves behind his Medal of Honor. Many viewers miss this important gesture.

- Some of the Navy SEALs in the film were played by real SEALS.

- While the film was shooting Alcatraz was still open to the public and many visitors watched the movie being shot, but in December of 1995 the federal government, which owns the island, partially shut it down due to stalled budget talks and filming continued with no visitors present.

- The coordinates given for Alcatraz when the missile is launched are 67°25'0"N by 37°25'0"W. These actually would put them in Greenland. (Alcatraz is actually at 37°49'36"N by 122°25'23"W.)

Reel Marines

❖ Throughout the movie, the Marines refer to each other as soldiers. A real Marine would *never* do that!

❖ Right before the Marines betray General Hummel, Captain Darrow refers to Gunnery Sergeant Crisp simply as "Sergeant." NCO's and SNCO's in the Marine Corps are always referred to by their full rank, i.e, "Gunnery Sergeant," or sometimes the more familiar "Gunny" - but never simply "Sergeant." No real Marine would ever make that error.

❖ When talking to General Hummel one of the characters says, "Let's be all we can be," which was the slogan for the Army at the time, not the Marines.

❖ Mason was supposed to have been captured in 1962 after hiding the microfilm, but at the end of the movie Goodspeed, while viewing it, asks his girlfriend if she wants to know who killed Kennedy - an event which occurred in November of 1963.

Reel Marines

THE SHORES OF TRIPOLI

> Release date: 11 March 1942
> Running time: 86 minutes
> Historical context: Peacetime/World War II
>
> Tagline: Romance... Comedy... Thrills... with Uncle Sam's fighting "Devil Dogs"!
>
> Cast
>
> John Payne - Chris Winters
> Randolph Scott - Sergeant Dixie Smith
> Maureen O'Hara - Mary Carter
>
> Quote: "I bet I'm the first Leatherneck in history that ever kissed a Lieutenant." – Private Chris Winters

Highlights

To the Shores of Tripoli was released just as World War II began for the United States, which gives the film a certain historical appeal above and apart from whatever entertainment value it may have. And it's certainly a product of its time - through and through, a recruitment film for the Marine Corps. Everything else is really secondary, with the focus of the story being the greatness of the Marines, and how any man worth his salt should join up and serve his country.

Reel Marines

Plot

Arrogant Chris Winters is ordered by his father, Captain Christopher Winters, to join the Marine Corps after he is expelled from Culver Military Academy for misbehavior. Disregarding his father's plans for his future, Chris intends to take a cushy desk job provided by the influential father of his girlfriend Helene after he finishes basic training. Captain Winters, however, has written to his old friend Sergeant Dixie Smith and asked him to toughen up Chris. Sergeant Smith becomes Chris' drill instructor, and takes an immediate dislike to the sarcastic recruit.

The night Chris arrives in San Diego he meets Mary Carter, and unaware that she is a Navy nurse and impressed by her beauty and spirit, tries half-successfully to romance her. Mary is frightened by her attraction to Chris however, and cuts short their evening. The next day Chris begins training along with fellow recruits Johnny Dent, Okay Jones, Mouthy and Butch. He easily masters the tasks assigned by Sergeant Smith, and helps the earnest but clumsy Johnny. In sequences filmed at the Marine Corps Recruit Depot in San Diego, Smith gives Winters an opportunity to demonstrate his leadership potential by drilling his platoon, and to Smith's amusement the Marines mock Winters and perform slapstick antics during the drill as he marches them away. Later, as Sergeant Smith is enjoying himself, the platoon marches back and performs a sequence of close order drill which is close to perfection. Smith is greatly surprised by their skill until he looks over the platoon and notices several recruits have black eyes, chipped teeth and bruises.

Chris attempts to pursue a relationship with Mary, but she reveals that as a nurse she holds rank equivalent to a lieutenant and cannot fraternize with enlisted men. Mary is

Reel Marines

troubled by Chris' lack of devotion to the military, but is still jealous when Helene appears on base one day and gets Chris to take her out. Later that evening when Chris returns he assures Mary that he cares only for her, and says if he takes the office job in Washington they can conduct their romance openly. Mary turns him down, and soon after Chris has more problems when he starts a fight with Dixie Smith, whom he accuses of bullying Johnny. They are both arrested for fighting, and despite the damage it could cause his career Smith states he started the fight so Chris will not get into trouble. Chris' barracks mates, angry that he caused Smith's demotion, snub him - and Chris decides to leave with Helene. Before they can leave the camp however, a field exercise is begun and Chris goes with the others to practice night maritime shelling.

As Sergeant Smith supervises his men in cleaning up floating targets he is knocked unconscious, and no one notices he is missing until they return to the ship. Despite the shelling Chris finds Smith, and rescues him just before the target he is on is destroyed. Later Chris says he risked his life only to erase the debt he owed Dixie for lying about the fight, and that he still intends to leave. Chris then asks Mary to accompany him, but she again replies that she belongs in the service. Chris departs with Helene, but while in the taxi they hear a radio report about Pearl Harbor. Helene declares that the report is a fabrication concocted by Orson Welles, but when Chris sees Dixie leading his Marines through the crowd, he realizes he truly is a Marine at heart and must join the fight. Chris dons his uniform as he marches with the other Marines, who are glad to see his change of heart. As they board a ship bound for war Chris notices that Mary is already aboard, and he then waves goodbye to his proud father, whose last words are, "Get a Jap for me."

Reel Marines

Trivia

- ❖ Titled after a lyric in the Marines' Hymn, which contains the phrase "...to the shores of Tripoli."

- ❖ The line, "To the shores of Tripoli" is a reference to the Battle of Derne in 1804, which was the first recorded land battle the United States fought overseas.

- ❖ The film is one of the last of the pre-Pearl Harbor service films to be produced.

- ❖ The Pearl Harbor attack occurred while the film was in post-production, causing the studio shoot a new ending where Payne re-enlists.

- ❖ *To the Shores of Tripoli* was dedicated to the 385 U.S. Marines who were then fighting at Wake Island.

- ❖ Randolph Scott also appeared as a Maine in *Gung Ho!*, where he portrayed "Colonel Thorwald."

- ❖ This was Maureen O'Hara's first film in Technicolor. She looked so good that she later earned the nickname "The Queen of Technicolor."

- ❖ This was Harry Morgan's film debut. He played Mouthy, and was credited as "Henry Morgan."

- ❖ Alan Hale Jr., who later gained fame as the "Skipper" on *Gilligan's Island,* plays Tom Hall.

- ❖ The exercise scene during Basic Training was filmed using actual Marines stationed at MCRD San Diego.

- ❖ In the film Dixie Smith is often addressed as "Sergeant" or even sometimes "Sarge," although he wears the three

Reel Marines

chevrons and two rockers of a Gunnery Sergeant. In the Marine Corps NCO's are always addressed by their full rank, so he should have been addressed as "Gunnery Sergeant," or if he allowed it, "Gunny."

❖ In several scenes Winters and the other members of his platoon are shown, while still in boot camp, wearing the Service "A" uniform complete with Marine Corps emblem. Recruits are not issued such a uniform until graduation, and are definitely not allowed to wear the eagle, globe and anchor because they are not yet considered Marines.

❖ Portions of *To the Shores of Tripoli* were filmed in Hawaii just before the Pearl Harbor attack, and according to studio publicity some of the cameramen managed to capture portions of the Japanese raid on film - although none of these scenes seem to have made their way into the final release print.

❖ Hugh Beaumont appears uncredited as an "Orderly." Best known for his work on TV's *Leave it to Beaver*, Beaumont played a Marine the *Salute to the Marines*, which was released the following year.

❖ Although this was a successful film, author Leon Uris ridicules it through the reaction of Marines who see it in his novel *Battle Cry*.

Reel Marines

THE SIEGE OF FIREBASE GLORIA

Release date: 27 January 1989
Running time: 95 minutes
Historical context: Vietnam

Tagline: Against all odds, they went to hell and back.

<u>Cast</u>

R. Lee Ermey - Sergeant Major Bill Hafner / Narrator
Wings Hauser - Corporal Joseph L. DiNardo
Robert Arevalo – NVA Colonel
Albert Popwell – First Sergeant Jones
Eric Hauser - Spider

Quote: "We are gonna refortify this shit hole and protect it like it was your daughter's cherry. I ain't gonna die here."
- Sergeant Major Hafner

Highlights

Over the years The Siege of Firebase Gloria has become something of a cult classic, and as a result copies are not as hard to find as they once were. The idea of a Sergeant Major leading a recon patrol doesn't make a whole lot of sense, but once you get past the rank issue it really doesn't make much of a difference what R. Lee Ermey is wearing on his collar. Besides, this movie is worth seeing for his "who do these heads belong to?" speech alone!

Reel Marines

Plot

The Siege of Firebase Gloria is a lean, straightforward film about a group of Marines defending an isolated outpost in Vietnam. While on a reconnaissance patrol Sergeant Major Hafner gets a glimmer of the impending Tet offensive when he comes across a massacred village. Later, when the patrol arrives at Firebase Gloria, Hafner takes command from the addled commanding officer and desperately tries to prepare the disorganized base for the coming assault. When the firebase is attacked the patrol remains to help defend it, and the Marines barely manage to hold off the Viet Cong. In the end they are forced to abandon it after losing too many men, and the Viet Cong commander discovers he is in a similar position. It was never his mission to win the battle, but to lead his men to their deaths in order to allow the North Vietnamese Army to take a more substantial role in the war.

There's nothing remarkable about the story, but after all this is a war movie - although the tale is told with efficient skill, and it's strikingly clear-eyed about the brutalities committed by both sides during the war. The Viet Cong are depicted with as much sympathy as the Americans, and in fact through smart juxtapositions the perspective of the Vietnamese soldiers becomes increasingly clear and compelling even though their dialogue is never translated. With clearly drawn yet realistic characters, a decent script, and effective camerawork, *Firebase Gloria* paints a much more vivid and unsettling portrait of the Vietnam experience than far more high-budgeted movies like *Full Metal Jacket*. Ermey gives a grounded, genuine performance, and straight-to-video star Wings Hauser is both scary and sad as a ruthless, alienated corporal.

Reel Marines

Trivia

- *Firebase Gloria* was filmed in the Philippines.

- Gerald Dwight Hauser, who was a star athlete in high school, earned the nickname "Wings" when he played wingback during his gridiron years.

- Erich Hauser, who plays a patrol member named Spider and is credited as Eric Hauser, is the brother of Wings.

- The script was co-written by William L. Nagle, who served with 3 Squadron (The Beagle Boys) of the Australian Special Air Service between 1965 and 1968.

- When Captain A.J. "Bugs" Moran shouts, "You picked a fine time to leave me, Lucille," he is referencing the song recorded by Kenny Rogers in 1977 - two years after the fall of Saigon and the end of the war.

- R. Lee Ermey wrote and/or improvised a great deal of his own dialogue. An example:

Hafner: (as he carries two severed American heads) Anyone know who these belong to? This is Corporal Miller (holds up head). He's dead. Hell, the whole gun crew's dead. And to add insult to injury, Charlie took the fifty-fucking caliber machine gun with him. I don't have any respect for Corporal Miller anymore, because he allowed his troops to relax. They let their guard down for five fucking minutes, and Charlie took advantage of it. Look at 'em, Goddammit! Pay attention. Stay alert! Stay alive! It's as simple as that!

Reel Marines

TO THE SHORES OF HELL

Release date: 16 March 1966
Running time: 82 minutes
Historical context: Vietnam

Tagline: The Hell-bustin' shoot-the-works epic of the U.S. Marines!

Cast

Marshall Thompson - Major Greg Donahue
Richard Arlen - Brigadier General F.W. Ramsgate
Robert Dornan - Dr. Gary Donahue
Bill Bierd - Gunnery Sergeant Bill Gabreski
Richard Jordahl - Father Jack Bourget

Quote: "When the fighting starts, the Marines are always there!" – Father Bourget

Highlights

This is what is known as a "campy classic." No one knew what Vietnam was going to evolve into at the time this movie was made, but in retrospect that fact certainly does add to the interest factor. Add to that a very "interesting" cast, and you have the makings of "Attack of the Killer Tomatoes" - Marine Corps style!

Plot

Marine Major Greg Donahue anxiously awaits orders to Vietnam because his brother Gary, a physician, has been captured by the Viet Cong. Donahue and his unit land in Da Nang, and there they learn Gary will be killed unless he treats wounded enemy soldiers.

Donahue receives permission to help his brother, and aided by Gunnery Sergeant Bill Gabreski, French priest Father Jacques Bourget, and Mic Phin, a Vietnamese guide, Donahue makes his way through the jungle to the abandoned French garrison where his brother is being held.

Gunny Gabreski is killed in an ambush, and although Mic Phin is seriously wounded by a boobytrap he still manages to enter the camp, help Gary escape, and take him to his brother and Bourget. Soon afterwards the party is attacked by guerrillas, but a Marine reconnaissance helicopter gives Donahue and the others enough support to put up a fight.

Later, with rescue only minutes away, the party is assaulted again, and both Bourget and Mic Phin are killed. At the last moment a helicopter arrives to extract the brothers, and they fly away while surveying the battlefield.

Trivia

❖ *To the Shores of Hell* represents one of the earliest efforts to portray the Vietnam War on film.

❖ Master Sergeant William V. Bierd was a Marine veteran of World War II, China, Korea, and Vietnam. Besides playing Gunny Gabreski, Bierd was an uncredited technical advisor on the film as well as a technical advisor on Marine Corps uniforms for *Gomer Pyle USMC*.

Reel Marines

- The part of prisoner of war Gary Donahue was played by former Congressman "B-1 Bob" Dornan, who was a member of the U.S. House of Representatives from Orange County, California from 1979-1996.

- The Marine Corps allowed Director Zens to film amphibious landing exercises at Camp Pendleton which appear at the start of the film.

- A Marine Corps HUS-1 helicopter was provided for the climax.

- Bob Dornan is the nephew of Jack Haley, who played the Tin Man in *The Wizard of Oz*.

- Dornan may possibly have co-written the film, because the writing credits list a "Robert McFadden" - and that is his Mother's maiden name.

- Richard Jordahl, who plays Father Bourget, said his performance "set Catholicism back fifty years."

- Thompson was a brother-in-law of actor Richard Long, best known for his role as Jarrod Barkley on ABC's *The Big Valley*. Thompson's wife, Barbara, was Long's sister.

- Thompson was a descendant of U.S. Supreme Court Chief Justice John Marshall, and was named after him.

- Thompson also produced, directed and starred in *A Yank in Viet-Nam* (1964), a film which could have had tragic consequences. It was an on-location anti-Viet Cong picture, and the VC put a price on his head during the shoot. The picture was important in that it was the first filmed during the war while under fire.

Reel Marines

TO THE SHORES OF IWO JIMA

Release date: 7 June 1945
Running time: 47 minutes
Historical context: World War II

Tagline: Courage has always been a virtue.

<u>Cast</u>

Harlon Block - Himself
John H. Bradley - Himself
Rene A. Gagnon - Himself
Ira H. Hayes - Himself
Franklin Sousley - Himself
Mike Strank - Himself
Franklin Delano Roosevelt - Himself
Richmond Kelly Turner - Himself

Quote: "With confidence in our armed forces, with the unbounding determination of our people, we will gain the inevitable triumph." – President Franklin Roosevelt

Highlights

If you want to see an authentic war film, this is the one. One look at the "cast" will tell you all that you need to know – as all of the "characters" are played by someone named "himself." Theatrical releases such as Sands of Iwo Jima and Flags of our Fathers have told this story well, but this is the film which captures the brutality of Iwo Jima.

Reel Marines

Plot

To the Shores of Iwo Jima is a 1945 Kodachrome color short war film produced by the Marine Corps which documents the Battle of Iwo Jima and provided the first opportunity for American audiences to see color the footage of the famous flag raising on Mount Suribachi.

The film follows the Marines through the battle in rough chronological order, from the bombardment of the island by Navy warships and carrier-based aircraft to the final breakdown of resistance, and after showing the taking of Suribachi it then switches to the footage of the second flag raising.

This film depicts with graphic energy the nearly month-long battle for Iwo Jima, a volcanic island southeast of Japan where twenty thousand Japanese and nearly seven thousand American troops were killed.

With all footage compiled by combat photographers from the Navy, Marines Corps and Coast Guard, the defending positions are softened by an extensive aerial and Naval bombardment followed by ten waves of landing craft occupied by men selected from the 110,000 who had to fight for every inch of black sandy soil. Only two hundred Japanese surrendered, with many being incinerated by flamethrowers, and all of the savagery is shown in dispiriting detail during the course of a work which was released only two months after the brutal engagement. The battle scenes are quite remarkable and shown in vivid color, and there are also several times when the mutated bodies of dead Americans are shown. There is also a scene where Marines are shown setting fire to a defensive position where several Japanese soldiers are located, and the film goes on to show the bodies of these men while they burn.

Reel Marines

Trivia

* Four cameramen, including Bill Genaust (who shot the famous flag raising sequence) were killed, and another ten were wounded, while filming this documentary.

* The film ends by acknowledging the thousands of Americans who had died in the month-long battle, and tells the audience that their deaths weren't in vain while showing a bomber aircraft taking off from the island for a mission over Japan.

* *To the Shores of Iwo Jima* is referenced on the television show *NCIS* in the episode *Call of Silence* (Season Two, Episode #7) in 2004. The NCIS team investigates a crime from WWII as a Medal of Honor recipient turns himself in for the murder of his best friend, and they review the film to recreate his actions. Charles Durning, who plays WWII vet Corporal Ernie Yost, is a WWII combat veteran in real life.

* Footage from *To the Shores of Iwo Jima* was edited into the John Wayne classic *Sands of Iwo Jima* in 1949.

* This film was edited by the aptly named Rex Steele of Warner Brothers.

* *To the Shores of Iwo Jima* was an Academy Award nominee for Best Documentary.

Reel Marines

Release date: 10 November 1970
Running time: 90 minutes
Historical context: Vietnam

Tagline: He's wanted by the U.S. Marines for AWOL, insubordination... and doing his thing!

Cast

Darren McGavin - Gunnery Sergeant Thomas Drake
Earl Holliman - Master Sergeant Frank DePayster
Jan-Michael Vincent - Private Adrian

Quote: "What kind of flower are you, boy? Are you a sweet pea, boy? Are you a sweet pea?" – Master Sergeant DePayster

Highlights

Tribes was a product of its times, having been made at the height of the Vietnam War. It is not the typical "boot camp" movie, in that it shows things from two very different perspectives and makes no judgment about which of them is better. I sometimes wonder what would have happened if Adrian had been allowed to graduate, and had been sent to Vietnam to fight. Would he have remained the same peaceful, nonviolent soul, or would the realities of "kill or be killed" combat have caused his survival instinct to override his penchant for peace and love?

Reel Marines

Plot

This drama tells the story of Private Adrian, a Vietnam War-era draftee who, despite being an anti-war "hippie," reluctantly reports to Marine Corps boot camp to fulfill his duty as an American. Adrian naturally excels as a leader, although his pacifist ideology presents continuing conflicts between himself and his Drill Instructor, Gunnery Sergeant Thomas Drake. Drake recognizes Adrian's leadership qualities but is conflicted as he grows to respect him, because he also realizes he represents everything Adrian opposes. At one point Adrian points out that his love of meditation is similar to Drake's drawing to relax, as he points to a sketch of a flying bird. Drake responds angrily, and denies he had drawn the picture.

Throughout training the Chief Drill Instructor, Master Sergeant DePayster, takes a dislike to Adrian. He repeatedly argues with Drake about him, and says the fact that the man is performing all of his assigned tasks is not enough. He considers Adrian's attitude grounds enough for him to be set back and placed in the Motivational Platoon, a disciplinary unit for problem recruits, but Drake disagrees and allows Adrian to graduate. DePayster then goes behind his back and files a complaint against them both with the Company Commander. Without Drake's approval the CO flunks Adrian and places him in "Moto" Platoon under DePayster, and when Drake accuses DePayster of carrying out a personal vendetta he replies, "I'll forget I heard that." Drake then takes the drawing of the bird from his desk drawer and hangs it up, thus signifying his own method of rebellion and freedom, and the platoon graduates without Adrian. The film ends as Drake awaits a new batch of recruits, with DePayster informing him that Adrian has deserted during the night.

Reel Marines

Trivia

- *Tribes* was originally broadcast as an ABC television *Movie of the Week*.

- The film was later released theatrically in Britain and Europe under the title *The Soldier Who Declared Peace*.

- Tribes first aired on November 10, 1970 – which is, of course, the Marine Corps' birthday.

- Somewhat surprisingly, and perhaps because it was aired on the Marine Corps Birthday, some recruits were allowed to watch this film while in boot camp.

- The movie was filmed on location at the Marine Corps Recruit Depot in San Diego, California.

- Some former Marines who saw this movie prior to leaving for boot camp have claimed to have tried some of the meditation techniques employed by Adrian "just to make it through the day."

- Jan-Michael Vincent played another Marine recruit with problems in *Baby Blue Marine*.

Reel Marines

UNCOMMON VALOR

Release date: 16 December 1983
Running time: 105 minutes
Historical context: Peacetime/Vietnam

Tagline: Seven men with one thing in common...

Cast

Gene Hackman - Colonel Jason Rhodes
Patrick Swayze - Kevin Scott
Fred Ward - Wilkes
Reb Brown - Blaster
Randall "Tex" Cobb - Sailor
Robert Stack - MacGregor

Quote: "Most human problems can be solved by a suitable application of high explosives." - Blaster

Highlights

Uncommon Valor follows a similar path to a great book of the 1980s, Mission MIA by J.C. Pollock. The central premise of leaving no man behind is the same, and both are important because those who go in harm's way must know they will not be forgotten. It has been said the Marine Corps will commit whatever forces are necessary to rescue a wounded Marine on the battlefield while under fire, and this story takes it to the next level when nearly a dozen men go back to a war which is long over for them.

Reel Marines

Plot

Uncommon Valor takes place in the early 1980s and is set in the context of the Vietnam War POW/MIA issue.

Retired Marine Colonel Jason Rhodes is obsessed with finding his son Frank, who has been listed as "Missing in Action" since 1972. Then, after ten years of searching Southeast Asia and turning up several leads, Rhodes comes to believe that Frank is still alive and being kept in Laos as a prisoner of war.

After petitioning the United States government for help, but receiving none, Colonel Rhodes brings together a disparate group of Vietnam War veterans, including some who were a part of Frank's platoon. There's Wilkes, a "tunnel rat" who now suffers from PTSD, "Blaster," a demolitions expert who believes most problems can be solved with a suitable application of high explosives, and "Sailor," a mental case with a heart of gold. In addition two helicopter pilots, Distinguished Flying Cross recipient Johnson (Harold Sylvester) and dysfunctional Charts (Tim Thomerson) round out the group. Former Force Recon Marine Kevin Scott also joins the team. Because Scott is young and has never served in combat he clashes with Sailor, but when Colonel Rhodes reveals that he is the son of a pilot who was shot down in Vietnam and listed as "MIA" the rest of the team accepts him immediately.

With the financial backing of rich oil businessman MacGregor, whose son served in Frank's platoon and is also listed as MIA, the men secretly train near Galveston, Texas before embarking on their trip to Laos in an attempt to bring back the POWs – but some unexpected obstacles have to be dealt with along the way which force Rhodes and his team to improvise and adapt.

Reel Marines

Trivia

❖ Gene Hackman, who plays Colonel Rhodes, served in the Marine Corps as a PFC from 1947 to 1954.

❖ The helicopters used in the film were purchased (as opposed to rented) and repainted, since the United States Department of Defense was unwilling to rent the production military-spec Huey or JetRanger helicopters due to the perceived anti-government nature of the film.

❖ The Laotian POW camp was built in the Lumahai Valley on the island of Kauai, Hawaii.

❖ During Colonel Rhodes' first conversation with Major Johnson (Harold Sylvester), the Major relates how he won the Distinguished Flying Cross in Vietnam. The story closely resembles a similar real life rescue from the book and movie *Bat*21*, which Gene Hackman starred in five years later.

❖ Gene Hackman asked Kris Kristofferson, his good friend and co-star in the 1972 movie *Cisco Pike*, if he wanted to do a cameo in this film. Kristofferson, a former Airborne Ranger, had to decline because he was on a concert tour.

❖ Randall "Tex" Cobb, who plays "Sailor," is convincing is his fight with Patrick Swayze for a reason. He fought Larry Holmes for the WBC World Heavyweight title in 1982, and although he suffered a defeat at the hands of Holmes, the bloody one-sidedness of the fight so horrified sportscaster Howard Cosell that he vowed never to cover another professional boxing match - which Cobb jokingly referred to as his "gift to the sport of boxing."

Reel Marines

WAKE ISLAND

Release date: 11 August 1942
Running time: 87 minutes
Historical context: World War II

Tagline: Tougher than leather... harder to kill! Daring Americans fighting against all odds!

Cast

Brian Donlevy - Major Geoffrey Caton
Macdonald Carey - Lieutenant Bruce Cameron
Robert Preston - Private Joe Doyle
William Bendix - Private Aloysius K. Randall
Walter Abel - Commander Roberts

Quote: "The Marines fought a great fight. They wrote history. But this is not the end. There are other leathernecks, other fighting Americans, a hundred and forty million of them, whose blood and sweat and fury will exact a just and terrible vengeance." - Narrator

Highlights

The story of Wake Island is one which ranks with those of the Alamo, Masada and the 300 Spartans, and it is surprising how good this movie is when you consider how soon after the actual event it was made. Many details of the battle were unknown until after the war when the captured Marines came home, and yet the writers still did well.

Reel Marines

Plot

Wake Island, a sandbar rising twenty-one feet out of the South Pacific and with a land mass of less than three square miles, was among the first U.S. outposts to be hit by the Japanese – in fact, the attack was virtually simultaneous with the one against Pearl Harbor.

Wake Island the movie was among Hollywood's earliest responses to America being attacked and drawn into World War II, and the Marines defending the island became instant war heroes, akin to the martyrs of the Alamo. Nothing could be done to rescue or even to reinforce and resupply them, but even so they fought on through air attacks and naval bombardments for two weeks until, badly outnumbered and outgunned, they were finally overrun.

The searing historical context had a lot to do with the movie's impact in 1942, and the sight of the dark forms of enemy planes coming over the horizon for the first time still carries a shock. *Wake Island* is a decent film, and it doesn't dishonor its subject with sham heroics and grandstanding - and most importantly, it helped to rally a Nation.

The first half hour sets up the allegory of America as melting pot - there's even a corporal named Goebbels - establishes horseplay as the coin of democratic discourse, especially for Gyrenes Robert Preston and the Oscar-nominated William Bendix, and fosters familiar friction between the new Commanding Officer played by Brian Donlevy and civilian construction supervisor Shad McClosky, portrayed by Albert Dekker.

William Bendix and Robert Preston as the two boisterous Marines provide the comic relief, but even they die valiantly in the end as the construction boss and his men die alongside the Marines while manning the last line of defense.

Reel Marines

Trivia

- ❖ The winner of four Oscar nominations, *Wake Island* was one of the first major Hollywood films to deal with America's forced participation in World War II.

- ❖ Paramount began work on this movie before the real life battle for Wake Island was even over.

- ❖ In a History Channel special called *Wake Island: The Alamo of the Pacific*, the survivors of the conflict called this movie one of the greatest works of fiction ever produced by Hollywood, mainly because it indicates there were no survivors.

- ❖ The end of the movie shows the Marines fighting to the bitter end, but in real life some of them were captured after surviving the first waves of the Japanese attack.

- ❖ In real life both the Marine and Naval commanders, Commander Winfield Cunningham and Major James Devereaux, were sent to POW camps and survived the war.

- ❖ Devereaux continued his career in the Marine Corps after being repatriated, retired as a Brigadier General, and was elected to Congress from his home state of Maryland.

- ❖ MacDonald Carey, who portrays Lieutenant Cameron, joined the Marines for real in 1943 and served for four years.

- ❖ Carey's character is based on the exploits of Captain Henry T. Elrod, who died on Wake and was awarded the Medal of Honor posthumously.

Reel Marines

❖ William Bendix, who plays Private Aloysius Randall, got "promoted" when he appeared as Marine Corporal Aloysius T. "Taxi" Potts in 1943's *Guadalcanal Diary*.

❖ The unfortunately named Goebbels (Philip Van Zandt) is constantly teased by the other Marines because he shares a name with the Nazi Minister of Propaganda.

❖ Many critics view the movie's portrayal of the Japanese as unfair and racist, but fail to remember the film was released only months after the attack on Pearl Harbor.

❖ In his book *Movie-Made America: A Cultural History of American Movies*, Robert Sklar states that this movie was "the first of many to dramatize American war heroics for the home front."

❖ The filming location was the Salton Sea in California.

❖ The seminal moment occurs when Donlevy, weary and battle-stained, is asked via radio what the defenders need, and replies to the American mainland with the defiant and legendary challenge, "Send us more Japs!"

❖ The credits inform us that the events depicted are "recorded as accurately and factually as possible," and then claims that none of the characters are actually based on people involved in the defense of Wake Island – which is, of course, ridiculous.

Reel Marines

WHAT PRICE GLORY?

Release date: 23 November 1926 / 25 July 1952
Running time: 116 minutes (1926) 111 minutes (1952)
Historical context: World War I

Tagline: From the Halls of Montezuma, to the Shores of Tripoli... they fought... and fought... and fought...

Cast (1926)

Edmund Lowe - 1st Sergeant Quirt
Victor McLaglen - Captain Flagg
Dolores del Río - Charmaine de la Cognac

Cast (1952)

James Cagney - Captain Flagg
Corinne Calvet - Charmaine
Dan Dailey - 1st Sergeant Quirt

Quote: "It's a lousy war, kid... but it's the only one we've got."
- Captain Flagg

Highlights

What Price Glory? was originally written as an anti-war play, but in typical Marine Corps fashion we have made it our own by embracing the combat and camaraderie, and ignoring the political undertones. The quality of the production is evident in the fact that it was remade by Hollywood legend John Ford, and spawned a series of "Flagg and Quirt" sequels.

Reel Marines

Plot

Flagg and Quirt are veteran Marine sergeants whose rivalry dates back a number of years. Flagg is commissioned as a Captain and placed in command of a company on the front lines of France during World War I, and when Sergeant Quirt is assigned to Flagg's unit as the senior non-commissioned officer they quickly resume their rivalry - which this time involves the affections of Charmaine, the daughter of the local innkeeper.

The first segment centers on the introduction of the central characters over the course of three years, starting in China where they serve as legation guards. Quirt is a hard-boiled non-commissioned officer of Marines, while Flagg is "soldiering for wages and loving, and fighting for fun." The next introduction is of Shanghai Mabel, who has "just divorced the Army and is announcing her engagement to the Marines." The involvement amongst Flagg, Quirt and Mabel then moves to the Philippines, followed by the next segment in 1914 France, and then the finale in 1917.

The two protagonists continue their private war as all hell breaks loose around them. When they aren't blowing the brains out of the Germans, Flagg and Quirt are vying for the attentions of Charmaine. The film alternates effectively between low comedy and grim melodrama, and reaches a dramatic high point when fatally wounded mamma's-boy Private Lewisohn screams, "Stop the blood! Stop the blood!" When the smoke clears, Flagg and Quirt decide to go over the hill for the sake of Charmaine, but when duty calls the two friendly enemies march shoulder to shoulder towards new adventures. The battle scenes were terrifyingly realistic for the time, but the most memorable aspect of the picture is the ribald byplay between Flagg and Quirt.

Reel Marines

Trivia

- The film *What Price Glory?* is based on a 1924 play by Maxwell Anderson and Laurence Stallings.

- *What Price Glory?* has been made twice as a film. The first version was released in 1926, and the second in 1952. Both versions follow the same general plot.

- In the argument between Sergeant Quirt and Captain Flagg in the silent version, the actors actually swore at each other. Hundreds of complaint letters were received from angry lip-readers who recognized the words.

- Actor Jack Fay died at age twenty-five on November 15, 1928 of injuries sustained in an explosion during the shooting of this film.

- In 1942 Lowe and McLaglen played two similar Marines in RKO's *Call Out the Marines*.

- The 1952 version uses almost no dialogue from the original play, and was originally intended to be a musical.

- Robert Wagner, who also played the part of a Marine in *Halls of Montezuma* and *In Love and War*, appeared as Private Lewisohn in the 1952 version.

- Barry Norton, who played Lewisohn in the original, appeared in the remake as a priest.

- Flagg and Quirt appeared in the sequels *The Cock-Eyed World* (1928), *Women of All Nations* (1931) and *Hot Pepper* (1932).

Reel Marines

WIND AND THE LION

Release date: 22 May 1975
Running time: 119 minutes
Historical context: Peacetime

Tagline: Between the wind and the lion is the woman. For her, half the world may go to war.

Cast

Sean Connery - Raisuli
Candace Bergen - Eden Perdicaris
Brian Keith - President Theodore Roosevelt
John Huston - Secretary of State John Hay
Steve Kanaly - Captain Jerome

Quote: "Captain Jerome, United States Marine Corps, and you sir, are my prisoner." - Captain Jerome

Highlights

This is an excellent movie, with great performances by Sean Connery and Candace Bergen, but it is two other things which really stand out. The former Marine who plays the part of Teddy Roosevelt is so believable that it seems as if the former President has been reincarnated, and the scene in which a company of Marines double-times through the streets of Tangier in Dress Blues (the fighting uniform of the day) with fixed bayonets and then assaults the palace is fantastic.

Reel Marines

Plot

1904 Morocco is the scene of conflict between the powers of Imperial Germany, France, and the British Empire, all of whom are trying to establish a sphere of influence in that country. Mulai Ahmed er Raisuli is the leader of a band of Berber insurrectionists opposed to Sultan Abdelaziz and his uncle, the Bashaw (Pasha) of Tangier, whom Raisuli considers to be corrupt and beholden to the Europeans. He kidnaps American Eden Pedecaris and her children, William and Jennifer, from their home after murdering Sir Joshua Smith, a British friend of Eden's. Raisuli then issues an outrageous ransom demand in a deliberate attempt to provoke an international incident, embarrass the Sultan, and start a civil war.

In the United States President Theodore Roosevelt is struggling for re-election and decides to use the kidnapping as both political propaganda (using the phrase "Pedecaris alive, or Raisuli dead!") and as a showcase to demonstrate America's military strength as a new power - despite the protests of his cautious Secretary of State, John Hay.

The American Consul to Tangier, Samuel Gummere, is unable to negotiate a peaceful return of the hostages, so Roosevelt sends the South Atlantic Squadron, under the command of Admiral French Ensor Chadwick, to Tangier with orders to either retrieve Pedecaris themselves or force the Sultan to accede to Raisuli's demands. Throughout the story, however, Roosevelt finds himself gaining more and more respect for Raisuli, thinking him an honorable man who just happens to be his enemy.

The Pedecaris' are kept as hostages by the Raisuli in the Rif, far from any potential rescuers, and although her children seem to admire Raisuli, Eden finds him to be "a brigand and a lout." The Pedecaris family attempts an escape, helped by one of Raisuli's men, but are betrayed and given to a gang of desert thieves. Luckily Raisuli is able to track them down, and he kills the kidnappers. He then reveals he has no intention of harming the Pedecaris' and is merely bluffing. Eden and Raisuli then seem to become enamored of each other as Raisuli tells the story of how he was once taken captive by his brother, the Bashaw, and kept in a dungeon for several years.

Gummere, Chadwick, and Marine Captain Jerome tire of the Sultan's deceit and the meddling of the European powers, and decide to engage in "military intervention" to force the Sultan to negotiate. Jerome's company of Marines, supported by a small detachment of sailors, double-times through the streets of Tangier with fixed bayonets and much to the surprise of the European legations overwhelms the Bashaw's palace guard and takes him prisoner.

The Bashaw finally agrees to accede to the Raisuli's demands, but during a hostage exchange Raisuli is betrayed and captured by German, French and Moroccan troops under the command of Von Roerkel, while Jerome and a small contingent of Marines secure the Pedecaris.' As Raisuli's friend, the Sherif of Wazan, organizes the Berber tribe for an attack on the Europeans and Moroccans, Eden attacks Jerome and convinces him and his men to rescue the Raisuli.

A three-way battle results in which the Berbers and Americans team to defeat the Germans, the French, and their Moroccan allies, and rescue Raisuli in the process. In the United States Roosevelt is cheered for this great victory, and the Pedecaris' arrive safely back in Tangier. Roosevelt later reads a letter he received from Raisuli, which says, "I, like the lion, must stay in my place, while you, like the wind, will never know yours."

Reel Marines

Trivia

- This film was based somewhat on the real-life Perdicaris incident of 1904, which involved the kidnapping of Ion Perdicaris, an American expatriate living in Tangier.

- The title was loosely inspired by a letter Raisuli sent to a Spanish military official in 1913. In it he stated, "You and I form the tempest. You are the furious wind. I am the sea." The Raisuli's letter to Roosevelt at the end of the film is a heavily phrased paraphrase of this.

- Steve Kanaly's character Captain Jerome was based on John Twiggs Myers, who served with distinction in the Spanish-American War, Philippine Insurrection, and most notably as commander of the American Legation during the Chinese Boxer Rebellion of 1900. Myers (then a Lieutenant) was commander of the Marine forces dispatched to Tangiers, but unlike the film's portrayal they were not engaged in combat against the Moroccan government. Myers would later become Commandant of the Marine Corps after World War I, retiring in 1934 as a Lieutenant General and dying eighteen years later.

- Director John Milius originally wanted Omar Sharif to play Raisuli and Faye Dunaway as Eden Pedecaris, but Sharif refused the part and Dunaway became ill and had to be replaced on short notice by Bergen. Anthony Quinn was also considered for Raisuli. Milius said he wrote the part of Eden with Julie Christie in mind, although she may not have actually been approached for the role.

- *The Wind and the Lion* gained considerable recognition in the Islamic world for its accurate, detailed, and sympathetic depiction of Berber and Islamic culture.

Reel Marines

- ❖ Brian Keith, who portrayed Theodore Roosevelt, served in combat as Marine during World War II. He went on to play a Marine colonel in the film *Death Before Dishonor* - which portrays the Islamic culture quite differently.

- ❖ Filming was done in Spain, with the towns of Seville, Almeria, and Madrid all doubling for Tangier and Fez, and the "Washington" scenes being filmed in and around Madrid.

- ❖ Director John Milius' primary source for the film was Barbara W. Tuchman's essay *Perdicaris Alive or Raisuli Dead!*, which was published in the August 1959 edition of *American Heritage* magazine. Much of the film's dialogue is taken verbatim from Tuchman's essay.

- ❖ The U.S. Marines and sailors used in the Tangier attack scene were Spanish special forces troops, along with about twenty USMC advisors, who marched with precision through the streets of Seville and Almeria en route to the Bashaw's palace. According to Milius, the Marine Corps actually shows this scene to its advanced infantry classes.

- ❖ Virtually all of the film's stunts were performed by Terry Leonard, who also has a minor part as Roosevelt's boxing opponent early in the film. Only four stunt men were used in the entire final battle scene.

- ❖ While filming that scene, Antoine Saint-John (who played Prussian calvary officer Von Roerkel) revealed himself to be terrified of horses, and would often hide somewhere on the set when his sword fight with Sean Connery was to be filmed.

Reel Marines

❖ Several the film's crew are cast in the movie, most notably cinematographer Billy Williams as the gun-shooting, white-suited Englishman in the opening scenes at the villa.

❖ Special effects supervisor Alex Weldon appears as Roosevelt's Secretary of War Elihu Root, and Milius himself cameos as the one-armed German officer who gives the Sultan his Maxim gun to test-fire ("Herr Sultan is displeased?").

Reel Marines

WINDTALKERS

> Release date: 14 June 2002
> Running time: 134 minutes
> Historical context: World War II
>
> Tagline: Honor was their code.
>
> **Cast**
>
> Nicolas Cage - Sergeant Joe Enders
> Adam Beach - Private Ben Yahzee
> Roger Willie - Private Charlie Whitehorse
> Peter Stormare - Gunnery Sergeant Hjelmstad
> Christian Slater - Sergeant Pete "Ox" Anderson
>
> Quote: "His name was Joe Enders, from South Philadelphia. He was a fierce warrior, a good Marine." - Ben Yahzee

Highlights

The Navajo Code Talkers were an important part of Marine Corps history, but because the program remained classified for many years it wasn't until Windtalkers was released that they were recognized by the public for their contributions to the war effort. The story is good, but the special effects and attention to detail in recreating World War II battlefields are outstanding – and the leadership principle which tells us that "mission accomplishment" comes before "troop welfare" has never been more clearly explained and understood.

Reel Marines

Plot

The film begins with then-Corporal Joe Enders and a platoon of his fellow Marines fighting Japanese forces on Guadalcanal in 1943. The outnumbered Marines are killed one by one, and as Enders mourns over the body of a friend a grenade explosion knocks him unconscious.

Enders is transported to a field hospital, and by mid-1944 has recovered from his physical wounds except for the hearing in one ear. He will be considered unfit for duty unless he can successfully complete a hearing test, and a sympathetic female pharmacist helps him cheat. Enders is then promoted to sergeant, and returned to active duty.

Enders receives a top priority assignment protecting Navajo code talker Ben Yahzee, and less-jaded Sergeant "Ox" Henderson receives a parallel assignment protecting Navajo Charlie Whitehorse. They are told that the code can not fall into enemy hands, and if their code talker is about to be captured they are to kill him to ensure the Japanese can't break the cipher.

The Marines land at Saipan under heavy fire, and it is Yahzee and Whitehorse's first experiences in combat. Yahzee is disgusted by all the death around him and shocked at how Enders mercilessly kills the Japanese, and although he never opens fire on the Japanese he and Whitehorse instead send coded messages to direct the battleship bombardment of their positions.

When the beachhead is secured the Marines advance further into Saipan, where their convoy comes under fire by American artillery. Yahzee's radio is destroyed in the initial salvo, and the convoy is unable to call off the bombardment. They attack to escape the shelling, and during the battle Private Nellie is killed trying to save a wounded man. A plan

Reel Marines

is devised to send Yahzee, disguised as a Japanese soldier, and Enders, who is posing as his prisoner, behind Japanese lines to use their radio. Yahzee is forced into killing the radioman, and is able to redirect the artillery to fire onto the Japanese positions. Enders is awarded the Silver Star for saving the lives of his Marines, and he gives the medal to Private Pappas to send to Private Nellie's wife back home.

The Marines then move to a Japanese village where they make camp as Yahzee is recalled to headquarters. The village is attacked, and during the fight a Marine's flamethrower tank explodes and Enders is forced to shoot him. Anderson is decapitated defending Whitehorse, who eventually gets taken by the Japanese, and after hesitating Enders throws a grenade which kills him along with the Japanese. Yahzee later returns and learns Enders had to kill Whitehorse to prevent him from falling into enemy hands, and in anger he attempts to shoot him, but is stopped.

The Japanese next ambush them near a minefield, but they are able to fight their way to the ridge, and when they arrive they see artillery which is firing down on American troops. Gunnery Sergeant Hjelmstad (Peter Stormare) is then killed by enemy fire, and Enders becomes the new squad leader.

Yahzee fearlessly attacks the Japanese, but carelessly loses the radio - which they need to call in air support. Both Yahzee and Enders are shot as they attempt to retrieve it, and when they become surrounded Yahzee tries to get Enders to shoot him to protect the code - but Enders refuses and carries him to safety. Yahzee then calls in air support which saves the column, but sees Enders was shot in the chest while carrying him, and tries to stop the blood loss. Enders stops him and says with his last breath that he didn't want to kill Whitehorse. The film then ends with Yahzee back home in Arizona, performing a Navajo ritual to pay his respects.

Reel Marines

Trivia

❖ To add authenticity, MGM bought some genuine WWII-vintage radios from Samuel M. Hevener, a collector from Ohio.

❖ The Naval Air Weapons Station at Point Mugu was used to film the Camp Tarawa portion of the film, which was the Marines' pre-battle embarkation point.

❖ Weapons coordinator Robert "Rock" Galotti amassed over 500 vintage WWII-era firing weapons and 700 rubber replica weapons for the film from private collectors and prop houses. Also featured moving across battlefields are vintage Sherman tanks, their smaller Stuart brethren, and Japanese Hago tanks.

❖ Prior to filming most of the principal cast joined a core group of sixty-two extras for boot camp, where they endured a week of rigorous military training as WWII Marines.

❖ The production received assistance from the Department of Defense, which made Kaneohe Marine Corps Base available for the actor's basic training.

❖ Under the tutelage of retired Marine Corps veteran Sergeant Major James Dever and his active-duty Marine instructors, the cast learned how to walk, talk and think like Marines.

❖ Adam Beach, who plays Ben Yahzee, appeared in *Flags of Our Fathers* as Ira Hayes.

Reel Marines

55 DAYS AT PEKING

Release date: 29 May 1963
Running time: 154 minutes
Historical context: Boxer Rebellion

Tagline: A handful of men and women held out against the frenzied hordes of bloodthirsty fanatics!

Cast

Charlton Heston - Major Matt Lewis
Ava Gardner - Baroness Natalie Ivanoff
David Niven - Sir Arthur Robinson
John Ireland - Sergeant Harry

Quote: "Remember, it's just the same here as anywhere else in the world. Everything has a price. So pay your money, and don't expect any free samples." - Major Lewis

Highlights

At the beginning of the twentieth century there were no motion picture cameras to capture history in the making such as we had during World War II, but 55 Days at Peking almost makes up for that. All Marines are taught about the Corps' part in the Boxer Rebellion during boot camp, but it's hard to imagine what it must have been like without seeing this picture. The only major omission, from a Marine perspective, was the absence of Dan Daly, who received the Medal of Honor for his actions there.

Reel Marines

Plot

55 Days at Peking is a dramatization of the Boxer Rebellion which took place in 1900 China. Fed up by foreign encroachment, Dowager Empress Tzu-Hsi uses the Boxer secret societies to attack the foreigners within China, culminating in the siege of the foreign legations' compounds in Peking (now Beijing). The film concentrates on the defense of the legations from the point of view of the foreign nationals, and the title refers to the length of the defense by the colonial powers of the legations district of Peking.

Foreign embassies in Peking are being held in a grip of terror, and the Boxers set about massacring Christians in an anti-Christian nationalistic fever as U. S. Marine Corps Major Matt Lewis heads a force of multinational soldiers and Marines tasked with defending the compound. Inside the besieged legation the British Ambassador then gathers the beleaguered representatives of a dozen governments into a defensive formation.

Included in the group of high-level dignitaries is the sultry Russian Baroness Natalie Ivanoff, who begins a romantic liaison with Lewis. As the group fortifies their positions and conserves food and water in an attempt to save some hungry children, they await the arrival of expected reinforcements - but the wily Empress is at the same time plotting with the Boxers to break the siege with the aid of Chinese troops.

Just when it seems that the Americans and Europeans are about to be overrun, the forces of the Eight-Nation Alliance arrive and lift the siege over the legations district and puts down the rebellion - an event which foreshadowed the demise of the Qing Dynasty.

Reel Marines

Trivia

- Director Nicholas Ray was best known for his 1955 film *Rebel Without a Cause* starring James Dean. Ray was a tortured individual at the time of the production of *55 Days at Peking* - somewhat akin to the Dean persona he helped to create for *Rebel*. Paid a very high salary by producer Samuel Bronston to direct *55 Days*, Ray had an inkling that taking on this project, a massive epic, would mean the end of him and that he would never again direct another film.

- Nicholas Ray also directed John Wayne and Robert Ryan in *Flying Leathernecks*.

- Due to mainland China's hostility and isolation from the Western world, a full-scale sixty-acre replication of circa 1900 Peking (sewers and all) was built in the plains outside Madrid.

- A number of costumes for the Royal Chinese Court (the Empress, Prince Tuan's, etc.) were authentic ones from Tzu-Hsi's actual court. They were loaned by an illustrious Florentine family (which wished to stay anonymous) who were able to rescue them from the collapse of the dynasty right after the Boxer rebellion.

- The film, which was shot in Spain, needed thousands of Chinese extras, and the company sent scouts throughout Spain to hire as many as they could find. As a result many Chinese restaurants in Spain closed for the duration of the filming because the staff - often including the restaurants' owners - were hired away by the film company. The company hired so many that for several

Reel Marines

months there was scarcely a Chinese restaurant to be found open in the entire country.

❖ In the scene where the Boxers are coming over the wall, there is a brief shot of U.S. Marines manning a water-cooled Browning M-1917 machinegun. That weapon was introduced in 1917 during WWI, hence its nomenclature, but this film is set in 1900. The Marines did in fact have and use machineguns during the siege, but they were Browning-Colt M-1895 machine guns - which were distinctly different from the gun shown.

❖ The rifle used by the Marines in the film is a .303 Lee Enfield British rifle. The U.S. .30-40 Krag-Jorgensen rifle would have been correct.

Reel Marines

Reel Marines

101 MORE FILMS

1969 (1988) Ralph and Scott (Kiefer Sutherland and Robert Downey Jr.) live in a small-minded town at the onset of wide public dissatisfaction with the Vietnam War. While Scott's brother enlists in the Marine Corps, he and Ralph are outspoken in their opposition to the war. Scott's attitude alienates him from his father, and he and Ralph leave town to enjoy their 'freedom.'

A Guy, a Gal, and a Pal (1945) Marine Jimmy Jones (Ross Hunter) is the "guy," the "gal" is Jimmy's sweetheart Helen Carter, and the "pal" is the couple's self-appointed chaperone, ten-year-old Butch. Helen's dilemma - should she marry Jimmy, or settle for financial security in the form of rich civilian Granville Breckinridge?

A Message To Garcia (1936) The real life incident on which the film is based involved Marine Lieutenant Rowan's (John Boles) trip to Cuba carrying a message to General Garcia from President McKinley that the United States was declaring war on Spain and was eager to have Garcia's cooperation.

A Yank in Vietnam (1964) In this Vietnam war drama a Marine (Marshall Thompson) survives a helicopter crash and lands in enemy territory. Rebels help guide him through the dense jungles to safety, and along the way he saves a POW and ends up falling in love with the man's daughter. Filmed in South Vietnam.

Reel Marines

Abroad With Two Yanks (1944) Biff and Jeff, two American Marines on furlough in Australia during the Second World War, are enjoying their time the way most leathernecks on leave do. When they meet the beautiful Joyce they both fall head over heels for her, and start competing for her attentions. As their time begins to run out, the schemes they each come up with to win her affection and foil the other's plans to do the same become more and more outrageous.

Aliens (1986) Fifty-seven years after her ordeal with an extraterrestrial creature, Ellen Ripley (Sigorney Weaver) is rescued by a deep salvage team during her hypersleep. When she discovers transmissions from a colony which has since settled on the alien planet suddenly stop, Ripley is offered a chance to team up with a group of "Colonial Space Marines" who descend to the planet and investigate the alien presence.

Baa Baa Black Sheep (1976) Pappy Boyington (Robert Conrad) is the squadron-leader of a group of fighter pilots stationed on an island in the Pacific during World War II. Pappy often needs to intercede in altercations at the base, but everyone seems to pull together when they are assigned missions in the air. Pilot for the TV series *Black Sheep Squadron*.

Baby Blue Marine (1976) A would-be Marine (Jan Michael Vincent) fails basic training, and is sent home wearing the "baby blue" fatigues of a washout. En route he is mugged by a battle-fatigued Marine Raider (Richard Gere), who leaves him to hitch-hike home in an undeserved hero's uniform. A small Colorado town takes him in, and treats him like the hero he appears to be.

Reel Marines

Band of the Hand (1986) In an attempt at resocialization, five hopeless juvenile criminals are sent away from prison and into the Everglades for survival training under a Native American Marine combat vet (Stephen Lang). Upon completion of the program, the group buys a vacant house in a dangerous part of Miami and slowly rebuilds the neighborhood, kicking out the pimps, prostitutes and drug dealers.

Battle Flame (1959) A group of five nurses is captured by North Korean troops and need to be rescued. At the head of the rescue task force is Marine Lieutenant Frank Davis (Scott Brady).

Battle Zone (1952) Danny (John Hodiak), a Marine Corps veteran of World War II, re-enlists when the Korean War breaks out. He joins a Marine motion picture unit specializing in combat footage and re-encounters Mitch, a former pal. Danny discovers Mitch is now engaged to the girl Danny left behind in Rome during the previous war, and a rivalry develops between them which spills over onto the battlefield until a secret mission behind North Korean lines brings their rivalry to the boiling point.

Beach Red (1967) U.S. Marines led by Captain MacDonald (Cornel Wilde) and Gunnery Sergeant Honeywell (Rip Torn) land on an unnamed Japanese-held Pacific island. The film then moves on, showing the Americans consolidating their gains, and looks back at the lives of some of the combatants - both American and Japanese.

Brothers (2009) Sam Cahill (Tobey Maguire) and Tommy Cahill (Jake Gyllenhaal) are siblings. A Marine about to embark on his fourth tour of duty, Sam is a steadfast family man married to his high school sweetheart Grace (Natalie Portman), with whom he has two young daughters. The film opens with Tommy being released from jail for armed robbery, not long before Sam departs for Afghanistan. Soon news comes that Sam's helicopter has crashed, killing all of the Marines aboard, but in reality he and hometown friend Private Joe Willis have been taken prisoner. With Sam "gone," Tommy develops a relationship with Grace and the kids. Sam is rescued and returns home, clearly traumatized by his experience, and confronts his brother.

Busses Roar (1942) A gang of Axis spies decide to use a passenger bus to secretly transport a bomb to a coastal oil field. The bomb is set to go off upon arrival and wipe out the passengers along with the oil deposits, but among the passengers is a Marine Sergeant named Ryan (Richard Travis), who senses something's amiss and races against time to save himself and the others.

Calling All Marines (1939) A young gangster joins an enemy espionage agency and agrees to enlist in the Marine Corps so he can pilfer plans for a newly developed aerial torpedo. To get in, the crook steals an innocent young man's papers and forges them for himself. Later he gets tossed into jail, but enemy agents help him escape - but by this time the gangster has decided not to betray his beloved country. He puts up a fight, manages to elude the spies, and returns to the Marines - where he works to bring the enemies to justice.

Call Out the Marines (1942) After several years' hiatus Curtis and McGinnis (Edmund Lowe and Victor McLaglen) rejoin the Marines as sergeants, and while stationed in San Diego duke it out over the attentions of a cabaret singer who turns out to be linked to a gang of enemy saboteurs.

Reel Marines

China Venture (1953) In late 1944 an American guerilla unit led by Marine Captain Matt Reardon (Edmond O'Brien) learns a Japanese plane carrying an Admiral has crashed in China, in warlord-held territory. Reardon and his men are sent on a mission to ransom the Admiral, treat his injuries, and bring him back to American lines.

Come On, Leathernecks (1938) Jimmy Butler (Richard Cromwell), the star-quarterback for the Naval Academy, has no intention of following his father, a career Marine officer, into the service following graduation. He has already signed a contract to play professional football, but Marine Lieutenant Henry Dolan, an aide to Colonel Butler, tricks Jimmy into going to the San Diego Marine Base to tell his father he plans to resign his commission. Colonel Butler is on a Philippine island in the pursuit of a gun-running gang of smugglers, and through more trickery by Dolan that is where the newly-commissioned Lieutenant ends up.

Come On Marines (1934) "Lucky" Davis (Richard Arlen), a ladies-man and devil-may-care Marine Sergeant, is leading a Marine-squadron on an expedition through a Philippine jungle where an outlaw bandit is leading a guerilla-war rebellion. Their assignment is to rescue a group of children from an island mission which has been cut off, and it comes as a bit of a surprise when Davis discovers the "children" are actually a group of beautiful young women.

Coming Home (1978) Left alone in Los Angeles when her gung-ho Marine husband Bob (Bruce Dern) heads to Vietnam in 1968, proper wife Sally Hyde ("Hanoi" Jane Fonda) decides to volunteer at the V.A. hospital where her new friend Vi (Penelope Milford) works. There she meets Luke Martin (Jon Voight), a former high-school classmate

245

Reel Marines

and Marine who has returned from 'Nam a bitter paraplegic. As their relationship grows, Sally sees the effect of the war on the troops after they come back, inspiring her to rethink her priorities.

Cover Up (1991) Mike Anderson (Dolph Lundgren) is a tough investigative journalist and former Marine who is on a dangerous foreign assignment and finds his life in jeopardy when he uncovers a deadly labyrinth of political intrigue which threatens the lives of thousands. Dispatched to investigate a mysterious attack on an overseas naval base, Anderson finds himself back on familiar ground. Instinct makes him question the official CIA explanation blaming an unknown terrorist group called Black October, and armed only with his combat training and determination to uncover the truth he sets out to expose a complex and dangerous political web.

Crazy Legs, All American (1953) All-American football star Elroy "Crazylegs" Hirsch plays himself in this rousing film biography. Beginning with his years in a mid-Wisconsin high school, the film traces Hirsch's multi-lettered career at the University of Wisconsin. After service in the Marine Corps Hirsch turns pro, and eventually joins the LA Rams.

Cuban Love Song (1931) A cocky Marine (Lawrence Tibbett) stationed in Havana devotes his attentions to a voluptuous Cuban peanut vendor, but he is the "love 'em and leave 'em" type and when World War One breaks out he drops Velez like a hot tamale and heads for Europe. Ten years later, Tibbett returns to Cuba and gets a big surprise.

Dangerous Minds (1995) A drama starring Michelle Pfeiffer and based on the autobiography *My Posse Don't Do Homework* by former Marine LouAnne Johnson, who took

Reel Marines

up a teaching position at Carlmont High School in Belmont, California, where most of her students were African-American and Hispanic teenagers from East Palo Alto, a then-unincorporated town at the opposite end of the district.

Devil Dogs of the Air (1935) James Cagney plays reckless stunt flyer Tommy O'Toole, who is encouraged to join the Marine Flying Corps by his old Brooklyn buddy Lieutenant William Brannigan (Pat O'Brien). An undeniably talented flyboy, Tommy is also brash, obnoxious and pugnacious, quickly earning the enmity of his fellow trainees. After nearly "washing out" of the Corps, Tommy is eventually brought into line by the combined efforts of Brannigan and the rest of the "devil dogs."

Dogfight (1991) In the fall of 1963 Eddie Birdlace (River Phoenix) is an eighteen-year-old Marine who is about to ship out with three of his buddies for a tour of duty in Vietnam. Planning a massive blowout for their last night in San Francisco, Eddie and his buddies set up a contest they call a "dog fight" where each man contributes fifty dollars to the pot - and whomever can bring the ugliest date for their meeting later that night wins the prize.

East L.A. Marine: The Untold True Story of Guy Gabaldon (2008) This is the true story of an extraordinary man named Guy Gabaldon. A U.S. Marine of Hispanic descent, he single-handedly captured over 1100 Japanese during the bloody fighting on Saipan in the summer of 1944 and became one of the legendary heroes of World War II.

Enemy Agents Meet Ellery Queen (1942) Famous detective Ellery Queen (William Gargen) is caught in the crossfire between Gestapo agents and United States Marines

Reel Marines

who are trying to locate precious jewels being smuggled from Egypt in a mummy case.

Fighter Attack (1953) This film is based upon the real-life last mission of its star, Sterling Hayden. Though he's flown enough missions to be sent home, squadron leader "Steve" insists on leading an offensive and is shot down behind enemy lines. Rescued by resistance fighters (Joy Page and J. Carroll Naish), Steve becomes the "inside man" for his squadron and lays the groundwork for the destruction of the German supply lines.

Fighting Devil Dogs (1943) Masked villain "The Lightning" seeks to conquer the world with his arsenal of advanced electrical weaponry. Opposing him are two Marines, Lieutenants Tom Grayson (Lee Powell) and Frank Corby. Grayson has a special reason to defeat The Lightning because he killed his father, but first they must discover the villain's true identity.

Flight (1929) U.S. Marines Jack Holt and Ralph Graves are pilots stationed in Nicaragua, and are required to fly their Curtis fighter-bombers on dangerous missions. The flight scenes, shot without the benefit of special effects or back projection, are truly awe-inspiring, and served as stock footage for countless Columbia films in future years.

From Headquarters (1929) Gutsy Marine Captain Slappy Smith (Monte Blue) is assigned to rescue a passel of tourists from the Central American jungle. While fulfilling his duties Smith falls in love with one of the unfortunate tourists, and this poses a problem for native gal Innocencia, to whom a drunken Smith had previously pledged eternal devotion. Much of the film is stolen by Guinn "Big Boy" Williams as the slow-witted Sergeant Wilmer.

Reel Marines

Girls of Pleasure Island (1953) In 1945 Roger Halyard is a stiff-upper-lipped British gentleman who lives peacefully on a South Pacific island with his three nubile, naive daughters until 1,500 Marines arrive to transform the island into an airbase. Despite the best efforts of Halyard, his housekeeper and Marine Colonel Reade, romance blossoms between the three girls and a trio of handsome leathernecks.

Hail the Conquering Hero (1944) Woodrow Lafayette Pershing Truesmith (Eddie Bracken) is a small town boy whose father, "Hinky Dinky" Truesmith, was a Marine who died a hero in World War I. Woodrow has been rejected by the Marine Corps due to his chronic hay fever, but rather than disappoint his mother he pretends to be fighting overseas in World War II while secretly working in a San Diego shipyard. After a chance encounter in a bar in which he buys a round of drinks for six Marines headed by Sergeant Heppelfinger (William Demarest), the group decides to return Woodrow to his home so his mother will not have to keep worrying about him. The Marines, to the chagrin of Woodrow, encourage the charade by loaning him their medals - but word leaks out, and when they step off the train the seemingly-harmless deception has escalated beyond control as the entire town turns out to greet its homegrown hero.

Heaven and Earth (1993) A Vietnamese village girl survives a life of suffering and hardship during and after the Vietnam War as a freedom fighter, a hustler, a young mother, a sometime prostitute, and eventually the wife of U.S. Marine Steve Butler (Tommy Lee Jones). This was the final movie in Oliver Stone's Vietnam trilogy.

Reel Marines

Heaven Knows, Mr. Allison (1957) Marine Corporal Allison (Robert Mitchum) finds himself washed ashore on a small island in the Pacific where the only other inhabitant is Sister Angela, a Catholic nun. Allison falls in love with Sister Angela, but theirs is a relationship that is not meant to be. Their situation soon changes when Japanese troops arrive on the island, but Allison's presence proves essential when the U.S. Navy attacks.

Hell in the Pacific (1968) A Marine pilot (Lee Marvin) and a Japanese naval officer (Toshirō Mifune) are marooned on an uninhabited Pacific island during WWII. In order to survive they must accept their differences and work together, despite their two countries being at war. Containing little dialogue, this film is not dubbed or sub-titled, and authentically portrays the difficulties the two characters had communicating.

Hell to Eternity (1960) True life story of Guy Gabaldon, a Los Angeles Hispanic boy raised in the 1930s by a Japanese-American foster family. During the war, as his foster parents are interned at a camp for Japanese Americans, Gabaldon's ability to speak Japanese helps him become a lone-operating Marine hero. During the bloody capture of the island of Saipan, he convinces hundreds of Japanese to surrender after their general commits suicide.

Here Come the Marines (1952) The Bowery Boys get drafted into the Marines, and on their first day in basic training their commanding officer discovers Sach's dad is an old war buddy of his, so he makes him a sergeant and puts him in charge. While on the drill field they discover the body of a dead Marine, and find a playing card on him which they

Reel Marines

trace to a local gambling house where they suspect the Marine was murdered.

High Crimes (2002) Claire Kubik (Ashley Judd) finds out her husband's name is actually Ron Chapman, and that he's a former Marine accused of murdering seven innocent civilians in El Salvador during a raid. He admits he was there and changed his identity to escape prosecution, but insists he's innocent - so Claire enlists the aid of Charles Grimes (Morgan Freeman), an ex-Army judge advocate with an axe to grind.

Hold Back the Night (1956) Captain Sam McKenzie (John Payne) is a tough commanding officer guiding the fighting withdrawal of a Marine platoon in the snowy hills of Korea. Payne always carries an unopened bottle of whiskey, which he regards as a good-luck charm, and a series of World War II flashbacks explain the riddle of the unconsumed liquor.

Hot Pepper (1933) Former Marines Quirt and Flagg (Victor McLaglen and Edmund Lowe) team up once again and get involved in a nightclub, but trouble ensues when they both fall in love with a feisty woman and begin fighting over her.

Iceland (1942) When the U.S. Marines land in Iceland during WWII Captain James Murfin (John Payne) wastes no time scoping out the local female population. He makes a casual pass at skating champ Sonja Henie, only to discover she has mistaken his attentions for a marriage proposal.

If I Had a Million (1932) Dying tycoon John Glidden is so dissatisfied with his relatives and associates that rather than will his money to any of them, he decides to give it away in million-dollar amounts to strangers picked from the city directory - including Marine Steve Gallagher (Gary Cooper).

Reel Marines

In Harm's Way (1956) This film recounts the lives of several naval officers (including John Wayne) based in Hawaii as the U.S. involvement in World War II begins. The title of the film comes from a quote from American Revolutionary naval hero John Paul Jones: "I wish to have no connection with any ship that does not sail fast, for I intend to go in harm's way." While primarily a Navy movie, this film is of interest because Marine Colonel Gregory (George Kennedy) and his battalion parachute into combat during a pivotal scene, and former Marine Hugh O'Brian appears in a small role as an Air Corps Major.

In Love and War (1958) Three young San Francisco residents (Robert Wagner, Jeffrey Hunter and real-life Marine veteran Bradford Dillman) sign up for the Marines at the outbreak of WWII. The film traces the progress of all three in the Pacific, and emphasizes the characters' individual strengths and shortcomings. One of the men is a gung-ho patriot, the second is a perennial goof-off, and the third hopes to prove his worth to his wealthy father.

Inchon (1981) The film depicts the Battle of Inchon during the Korean War. The protagonist of the film is General Douglas MacArthur (Laurence Olivier), who led the surprise amphibious landing at Inchon in 1950 which was spearheaded by the 1st Marine Division. A subplot involves Marine Major Frank Hallsworth and his wife (Ben Gazzara and Jacqueline Bisset) who have problems in the mist of the war. Produced by the Reverend Sung Myung Moon's Unification Church.

Island of Desire (1952) Marine Corporal Michael "Chicken" Dolan (Tab Hunter) and nurse Elizabeth Smythe are the only survivors when a hospital ship hits a mine in the South

Reel Marines

Pacific during World War II. The two spend months alone on a deserted island and fall in love, but their peaceful existence is shattered when an airplane crashes on the island and the only survivor also falls for Elizabeth.

Join the Marines (1937) New York City cop Phil Donlan (Paul Kelly) leaves the department to join the U.S. Olympic team. He falls for the spoiled daughter of a Marine colonel, and winds up getting kicked off the team. He then joins the Marines to win the Colonel's approval, and everybody winds up on a jungle island fighting an outbreak of bubonic plague and putting down a native rebellion.

Leathernecking (1930) In this musical a Marine private falls in love with a socialite and is willing to do anything to win her affections, even if it means stealing his captain's uniform and posing as an officer. When that doesn't work he tries faking a shipwreck, but that goes awry and turns into a real one.

Let It Rain (1927) Sergeant "Let-It-Rain" Riley (Douglas MacLean) is a devil-may-care Marine who falls in love with a girl who he assumes to be rich. His rival for the girl's affection is his pal, Kelly. The guys find out that the object of their affections is but a modest switchboard operator, but she proves to be invaluable when she deciphers a code and discovers that a mail train is about to be robbed.

Letters From Iwo Jima (2006) Portrays the Battle of Iwo Jima from the perspective of the Japanese soldiers, and is a companion piece to Clint Eastwood's *Flags of Our Fathers*, which depicts the same battle from the American viewpoint. Interesting in that when Marines appear onscreen, they are the enemy.

Reel Marines

Maria's Lovers (1984) Marine Ivan Bibic (John Savage) returns to his Pittsburgh suburb after surviving a Japanese POW camp, and experiences regular nightmares. All the time he remained faithfully devoted to his childhood love, fellow ethnic Yugoslavian virgin Maria Bosic (Nastassja Kinski).

Marine Battleground (1966) An American nurse in Vietnam tells her story to a war correspondent (Jock Mahoney). The film chronicles her childhood in Korea during the 1950s when, during the battle at Inchon, her mother was killed and she was taken in by helpful Marines. When all but two of the leathernecks are killed, the girl vows to become a nurse to honor their memory.

Marine Raiders (1943) After Guadalcanal Major Steven Lockhard and Captain Dan Craig (Pat O'Brien and Robert Ryan) go to Australia, where one of the officers falls in love with a woman and the lovers wed just before he goes off to battle.

Marines, Let's Go (1961) A tale which follows Marine PFCs Skip Roth, Dave Chatfield and Desmond "Let's Go" McCaffrey (Tom Tryon, David Hedison and Tom Reese) on shore leave in Japan and at war in Korea.

Monkey On My Back (1957) The true story of welterweight boxing champion Barney Ross (Cameron Mitchell), whose meteoric ring career is interrupted when he joins the Marines at the onset of WWII. A highly decorated hero, Ross contracts malaria oversees and is given morphine for the pain - and by the time he returns to the States, he is a confirmed drug addict.

Reel Marines

Moran of the Marines (1928) Soon after joining the Marines Mike Moran (Richard Dix) is court-martialed for kissing a general's daughter ((Ruth Elder). His unit is then ordered to China, Elder is captured by bandits, and Moran rescues her. Silent.

No Leave, No Love (1946) In this WWII musical young hero Sergeant Michael Hanlon (Van Johnson) and his buddy decide to celebrate his receiving the Medal of Honor by partying. He is terribly excited about seeing his fiancée again, but unfortunately she has fallen for another man in his absence.

None But the Brave (1965) American Marines and Japanese soldiers stranded on a tiny Pacific island during World War II make a temporary truce and cooperate to survive various tribulations. The story is told through the eyes of the American and Japanese unit commanders, who must deal with an atmosphere of growing distrust and tension between their men.

One Kill (2000) Divorced Marine Captain Mary Jane O'Malley (Anne Heche) begins an affair with Major Nelson Gray (Sam Shepard) in this fact-based drama. Things turn nasty when she discovers he is married, and when he breaks into her house she pulls a weapon from beneath her pillow and kills him. The movie then moves to the courtroom.

Operation Secret (1952) This mystery centers on an American Marine (Cornel Wilde) who heroically served as a member of the French Foreign Legion and is falsely charged with the murder of a Resistance leader. Loosely based on the true story of Lieutenant Colonel Peter Ortiz.

Reel Marines

Professional Soldier (1935) Marine adventurer-for-hire Michael Donovan (Victor McLaglen) is hired by a political faction in a mythical European kingdom to kidnap a young prince so his employers can take over the government. In the end, when he realizes the political party he'd been working for plans to kill the prince and set up a dictatorship, he rescues the young monarch.

Red White and You (2008) Marine Recruiter Keith Kenworthy (John Moran) is the best in history. When he receives an award an ex-lover tracks him down and tells him he has a kid - and he just recruited him. (Editor's note: if there was a "technical advisor" supervising the uniforms worn in this short film, he should be hunted down and shot!).

Running Brave (1983) Dramatized true story of Marine Billy Mills (Robby Benson), a Native American who thrilled the world when he came from behind to win the 10,000 meter run in a huge upset at the 1964 Tokyo Olympics.

Sadie Thompson (1928) Sadie (Gloria Swanson), a woman of loose morals and sordid reputation, travels to the South Seas seeking a new life. She makes little effort to curb her hedonism when she's "entertaining" a group of U.S. Marines stationed on Pago Pago, but falls genuinely in love for the first time with Sergeant Tim O'Hara (Raoul Walsh).

South Pacific (1958) Film version of the Rodgers & Hammerstein Pulitzer Prize winning musical in which Marine Lieutenant Joseph Cable (John Kerr) and wealthy French planter Emile De Becque go on a reconnaissance mission against the Japanese.

South Sea Woman (1953) Marine James O'Hearn (Burt Lancaster), facing a court-martial for desertion, recalls how

Reel Marines

he was stranded in Shanghai. He and fellow Marine Davis White try to make it back to Pearl Harbor, but undergo several hair-raising adventures along the way - including involvement with a group of French resistance fighters.

Stars and Stripes Forever (1952) Film biography of the "March King," composer John Philip Sousa (Clifton Webb), from his early days with the Marine Corps Band through the Spanish-American War in 1898.

Star Spangled Banner (1917) An American widow living in London with her son Roger marries a Marine Colonel (Herbert Evans). Disappointed that he isn't British, the boy refuses to salute the American flag after the family relocates to the U.S. - but the day comes when Roger injures himself on a hunting trip, and when he doesn't come home the Colonel rounds up a group of his fellow Marines. They rescue him, and he finally sees the error of his ways. Silent.

Tarawa Beachhead (1958) Marine Sergeant Tom Sloan sees his CO break the law, and finds himself facing a dilemma. He wants to report it, but fears his superior officers will not believe him. As the battles rage on, the tension between the two mounts.

Taxi Driver (1976) Insomniac Marine Vietnam veteran Travis Bickle (Robert De Niro) works the nightshift, driving his cab throughout decaying mid-'70s New York City. Chronically alone, Travis cannot connect with anyone, not even the other drivers. He becomes infatuated with vapid blonde presidential campaign worker Betsy (Cybill Shepherd), and begins to condition himself for his imagined destiny, a mission which mutates from assassinating Betsy's candidate to violently "saving" teen hooker Iris (Jodie Foster) from her pimp (real-life Marine vet Harvey Keitel).

Reel Marines

The Bob Mathias Story (1954) Olympic decathalon winner Bob Mathias plays himself in this biopic. The film follows Mathias through his years at Stanford University, the 1952 Helsinki Olympics, and his time in the Marine Corps.

The Cockeyed World (1929) Marines Flagg and Quirt (Edmund Lowe and Victor McLaglen) fight over a sexy Russian girl, then a pretty blonde in Brooklyn, and finally wind up in a South American country where they fight for the favors of a beautiful senorita and try to put down a rebellion by the locals.

The Fighting Marines (1935) When the Marine Corps starts building a landing strip on Halfway Island in the Pacific Ocean they interfere with the secret hideout of masked villain "Tiger Shark," who begins to sabotage their efforts. Sergeant Schiller is abducted by the villain after developing a gyrocompass that could pinpoint his location, so Corporal Lawrence and Sergeant McGowan attempt to rescue him and stop the Tiger Shark.

The Godfather (1972) Vito Corleone is the aging head of a Mafia family. His youngest son Michael, a decorated Marine Captain, has returned from WWII just in time to attend the wedding of his sister. All of Michael's family is involved with the mob, but he just wants to live a normal life - until the attempted assassination of his father.

The Happiest Millionaire (1967) The (mostly) true story of eccentric Philadelphia millionaire Anthony J. Drexel Biddle (Fred MacMurray), who was an officer in the Marine Corps and an expert in close-quarters fighting. The last live-action film personally supervised by Walt Disney.

Reel Marines

The Last Warrior (1989) During the final days of World War II an American Marine (Gary Graham) finds himself on an otherwise deserted island with a Japanese soldier and a novice nun. Also released as *The Coastwatcher*.

The Leatherneck (1928) At the headquarters of the 6th Marine regiment in China, Calhoun (William Boyd) and Schmidt (Alan Hale) are facing court-martial for desertion. In a series of flashbacks, the viewer is apprised of the reasons for the two leathernecks' supposed dereliction of duty. Essential to the action are a third Marine, the unfortunate Joe Hanlon, and a mysterious Russian girl named Tanya.

The Leathernecks Have Landed (1936) An adventure yarn revolving around three boisterous Marines. Woody (Lew Ayres) is the headstrong one, Mac (James Ellison) the sincere one, and Tubby (Maynard Holmes) the roly-poly comic relief. Holmes is killed in a nightclub brawl for which Ayres gets the blame, but the real murderers are smugglers - so the disgraced Ayres joins the gang to bring them to justice.

The Marines Are Coming (1934) Lieutenant William "Wild Bill" Traylor (William Haines) is a wise-guy Marine, while Captain Ned Benton (Conrad Nagel) is his serious best friend. Both men vie for the attentions of a cute blonde, but a south-of-the-border tootsie complicates matters. Everything is resolved after an exciting battle between the Marines and a gang of Mexican bandits.

The Marines Are Here (1938) Jonesy (Gordon Oliver) is a self-centered character who comes to respect the Corps and everything it stands for under the less-than-gentle tutelage of Sergeant Gibbons (Guinn Williams). Jonesy proves he's

Reel Marines

truly one of the "few good men" during a battle between the Marines and a gang of bandidos.

The Marines Come Through (1943) Privates "Singapore" Stebbins (Wallace Ford) and "Junior" Murray (Grant Withers) are a pair of over-age leathernecks stationed in the South Seas. They battle over the affections of the fetching Toby Wing, and also foil a plan hatched by espionage agents to steal a revolutionary new bombsight.

The Marines Fly High (1940) When a platoon of Marines needs a safe place to stay in the Central American jungle, plantation owner Joan Grant (Lucille Ball) allows them inside her home. After she's kidnapped by a gang of bandits, the troop of Marines sets off to save her.

The Nun and the Sergeant (1962) During the Korean War a grizzled Marine Sergeant (Robert Webber) must undertake a potentially suicidal mission. He decides not to waste his best troops, and instead chooses from amongst his very worst and attempts to train them. His methods are harsh, his men hate him, and as soon as they cross into enemy territory they meet an injured American nun and her schoolgirls at their objective.

The Princess and the Marine (2001) Chronicles the true story of the star-crossed romance between Bahrainian princess Meriam Al-Khalifa (Marisol Nichols) and U.S. Marine Jason Johnson (Mark-Paul Gosselaar). Even if Meriam were not already committed to an arranged marriage, her strict Muslim parents would never approve of her union with an American Mormon. Determined to be together, the couple manages to sneak out of Meriam's homeland with forged papers.

Reel Marines

The Proud and Profane (1956) On Noumea in 1943 Lee Ashley (Deborah Kerr), the widow of a Para-Marine Lieutenant killed in the Battle of Bloody Ridge on Guadalcanal, joins the American Red Cross and falls for Lieutenant Colonel Colin Black (William Holden), the commander of a Marine Raider battalion.

The Rebel (1985) Chronicles the exploits of a nightclub singer (Debbie Byrne) and a Marine (Matt Dillon) in Australia. She works in a Quonset hut which was turned into a saloon, and helps stage shows for battle-weary troops returning from the South Seas. There she meets the Marine sergeant, who has just sailed in from Guadalcanal.

The Right Stuff (1983) Film adapted from Tom Wolfe's 1979 book about test pilots who were involved in high-speed aeronautical research at Edwards Air Force Base, as well as the original seven astronauts selected for Project Mercury. Of interest due to Ed Harris' portrayal of Marine Colonel John Glenn, the first man to orbit the Earth.

The Singing Marine (1937) Chronicles the exploits of young recruit Robert Brent (Dick Powell) who wins a radio contest and becomes an overnight singing sensation. It was from this musical that the Marine Corps got its signature anthem, *The Song of the Marines*.

The Steel Claw (1961) Marine Captain John Larsen (George Montgomery), who has lost a hand in an accident and has replaced it with a steel prosthetic hook, has to go behind enemy lines in the Philippines to rescue a general from Japanese captivity. Larsen discovers that the general has died, and it is almost impossible to get back to his rendezvous point.

Reel Marines

The Unbeliever (1918) Wealthy young American Philip Landicott (Raymond McKee), bred to class distinction and racial intolerance, enters the Marines during the First World War. In the course of his training and experiences in the trenches he comes to recognize the equality and brotherhood of men. Several genuine Marine officers, including future Commandant Thomas Holcomb, appear as themselves, as did members of the Third Battalion, Sixth Marine Regiment.

The Walking Dead (1995) Four black Marines are the only survivors of a platoon that's been decimated after landing behind enemy lines. Sergeant Barkley (Joe Morton) is a no-nonsense, Bible-quoting preacher, Cole Evans (Allen Payne) is an intellectual who is highly political, Joe Brooks (Vonte Sweet) is a cheerful, naïve, and brave young Marine, and Hoover Branche (Eddie Griffin) is a dope-smoking, foul-mouthed rebel who hates the war. Flashbacks reveal why each of them joined the Marines.

Till the End of Time (1946) The story concentrates on Marine veterans Cliff Harper (Guy Madison), Bill Tabeshaw (Robert Mitchum) and Perry Kincheloe (Bill Williams). Harper falls in love with emotionally distraught war widow Pat Ruscomb, Tabeshaw endures one disappointment after another as he tries to buy his own ranch, and Kincheloe, rendered legless by the war, intends to spend the rest of his life wallowing in self-pity. All three men find a new lease on life when they engage in a cathartic barroom brawl against a bigoted group of self-styled patriots.

Tripoli (1950) 19th-century clash between U.S. Marines and the pirates of Tripoli, with John Payne as Marine Lieutenant Presley O'Bannon. Howard Da Silva as Captain Demetrios, leader of a band of mercenaries, and Maureen O'Hara

Reel Marines

appears as Countess D'Arneau, who has come to Tripoli hoping to wed a local prince. After a great deal of byplay between the three stars, the action comes thick and fast as the Marines and the pirates "have at" each other.

True Lies (1994) Harry Tasker (Arnold Schwarzenegger) is an agent of Omega Sector, an organization which hunts down nuclear threats to the United States. When a middle-eastern terrorist network called "Crimson Jihad" threatens to nuke the U.S., Harry must stop them. While in pursuit of the terrorists, Marine AV-8B Harriers from Marine Attack Squadron 223 (VMA-223), nicknamed the "Bulldogs," take out a section of the Overseas Highway to the Florida Keys with missiles, and Arnold says, "Good shooting, Marines!"

Until They Sail (1957) Unemotional, "been-there-done-that" Marine Captain Jack Harding (Paul Newman) is assigned to investigate requests to marry local girls in World-War-II New Zealand. Harding begins to warm up when he meets war widow Barbara Forbes (Jean Simmons), a woman with three sisters who have their own entanglements.

We Are the Marines (1942) Deals with the history of the Corps from colonial times to 1942. The film's midsection details the arduous training at Parris Island and elsewhere, and wartime newsreel footage is adroitly blended with dramatized re-enactments to illustrate the contributions by and utter necessity for the Marines in World War II.

Without Reservations (1946) Kit Madden (Claudette Colbert), a best-selling novelist heading to Hollywood by train to oversee the film version of her latest novel, chooses Marine war hero Rusty Thomas (John Wayne) as the man who should portray her protagonist. Unaware of Kit's true identity, Rusty and his pal Lieutenant Dink Watson (Don DeFore) rail against the factual errors in her book.

World Trade Center (2006) In the aftermath of the September 11th 2001 terrorist attack on the World Trade Center, the building collapses on a rescue team from the Port Authority Police Department - and Officer Will Jimeno and Sergeant John McLoughlin are later found alive under the wreckage by former Marines Dave Karnes (Michael Shannon) and Jason Thomas (William Mapother). The Marine identified only as "Sergeant Thomas" in the movie's credits has since been identified as Jason Thomas, a former Marine who grabbed his uniform and immediately drove to Ground Zero to help. Because his identity was unknown at the time of filming Thomas, who is black, was portrayed by a white actor. The producers have since apologized for the miscasting, but Thomas laughed it off and told them not to be concerned, stating, "I don't want to shed any negativity on what they were trying to show."

Women of All Nations (1931) Pugnacious, girl-crazy Marine Sergeants Flagg and Quirt (Victor McLaglen and Edmund Lowe), the eternally bickering buddies first introduced in *What Price Glory?*, hopscotch from Panama to Sweden to Nicaragua to Turkey, battle over the affections of icy blonde Elsa, and find themselves in the middle of a sheik's harem.

PHOTO GALLERY

When people talk about Marines being portrayed in the movies the first name that usually comes to mind is John Wayne due to his memorable Oscar-nominated portrayal of Sergeant John Stryker in 1949's *The Sands of Iwo Jima*.

Reel Marines

PFC Al Thomas (Forrest Tucker) and Sergeant John M. Stryker (John Wayne) prepare to hit the beach in *Sands of Iwo Jima*.

John Agar, John Wayne and Forrest Tucker in action.

Reel Marines

John Wayne's role of Major Kirby in *Flying Leathernecks* was inspired by real World War II Marine ace Major John L. Smith.

Director Nicholas Ray chose Marine veteran Robert Ryan to play opposite John Wayne because he had been a boxer in college and was the only actor Ray knew who could "kick Wayne's ass."

Reel Marines

The one and only R. Lee Ermey as Gunny Hartman in *Full Metal Jacket*, asking Private Pyle about the caloric content of a jelly donut!

Aldo Ray and Tab Hunter having a heart-to-heart in 1955's *Battle Cry*, which was based on the bestseller by Marine veteran Leon Uris.

Reel Marines

Lieutenant Colonel Bull Meacham (Robert Duvall) reading his "Hogs" the riot act in *The Great Santini*.

Reginald Gardiner, Richard Widmark and Jack Webb (six years before he was T/Sgt Moore in *The D.I.*) in *Halls of Montezuma*.

Reel Marines

Kevin Bacon (R) as JAG prosecutor "Smiling Jack" Ross in *A Few Good Men*. He would go on to play a Marine LtCol a few years later.

Jack Nicholson as Colonel Jessup, telling Tom Cruise he "can't handle the truth!"

Kiefer Sutherland as Lieutenant Kendrick in *A Few Good Men*.

Reel Marines

Diane Keaton (as Kay Adams) and Al Pacino (as Captain Michael Corleone) in the Academy Award-winning classic *The Godfather*.

Ed Harris as Colonel John Glenn aboard *Friendship 7* in *The Right Stuff* (L), and Brad Davis as Phillip Caputo in *A Rumor of War*.

Reel Marines

Corporal DiNardo (Wings Hauser) was R. Lee Ermey's right-hand man in *The Siege of Firebase Gloria.*

Mort Sahl, Bradford Dillman (a real-life Marine veteran), Frances Nuyen, Steve Gant, Hope Lange, Robert Wagner, Dana Wynter, and Jeffrey Hunter in the 1958 film *In Love and War.*

Reel Marines

Alan Ladd, Sidney Poitier and Paul Richards in the groundbreaking 1960 Korean War epic *All the Young Men.*

Mickey Rooney, real-life Marine Corps veteran Hugh O'Brian and James Mitchum starred in *Ambush Bay.*

Reel Marines

Clint Eastwood as Gunny Tom Highway in *Heartbreak Ridge*, politely asking Corporal "Stitch" Jones where his bunk is.

Gunny Highway points out to kiss-ass Staff Sergeant Webster (Moses Gunn) that he couldn't build a good case of hemorrhoids.

Reel Marines

Then-unknown actor Richard Gere appeared as a Marine Raider in 1976's *Baby Blue Marine* long before he fell into Gunnery Sergeant Foley's clutches in *An Officer and a Gentleman*.

Jan-Michael Vincent starred in *Baby Blue Marine* six years after playing recruit Private Adrian in *Tribes*.

Reel Marines

Jack Webb as T/Sgt Jim Moore in *The DI*, telling Private Owens he should have let that poor sand flea eat all it wanted.

Cast of *Black Sheep Squadron*. Robert Conrad is wearing the skirt.

Reel Marines

Robert Mitchum (L) on the set of *Heaven Knows, Mr. Allison.*

Corporal Allison (Robert Mitchum) and Catholic nun Sister Angela (Deborah Kerr) are allies on a Japanese held island.

Reel Marines

Real-life Marine veteran Lee Marvin says hello to Japanese officer Toshirō Mifune in *Hell in the Pacific.*

Captains Steven Hiller and Jimmy Wilder (Will Smith and Harry Connick Jr.) get ready to kick some alien butt in *Independence Day.*

Reel Marines

Jake Gyllenhaal (as Anthony 'Swoff' Swofford) and Peter Sarsgaard (as Troy) in *Jarhead.*

John Garfield stars as real-life Guadalcanal hero Al Schmid in *Pride of the Marines.*

Reel Marines

Albert Dekker (as Shad McClosky), Brian Donlevy (as Major Caton), Walter Abel (as Cmdr Roberts), Macdonald Carey (as Lt. Cameron), and Robert Preston (as Pvt Joe Doyle) in *Wake Island*.

Robert Preston and Brian Donlevy man a machinegun and try to stop the Japanese onslaught in *Wake Island*.

Reel Marines

Major Frank Hallsworth (Ben Gazzara) with Jacqueline Bisset as his wife in *Inchon*.

Ben Gazzara and Richard Roundtree in the Reverend Sung Myung Moon's $100 million dollar turkey *Inchon*.

Reel Marines

It looks like Major Payne's Agent Orange is beginning to act up!

Damon Wayans as Major Benson Winifred Payne in *Major Payne*.

Reel Marines

Louis Gossett Jr. as Drill Instructor Gunnery Sergeant Foley in *An Officer and a Gentleman.*

"Sound off like you've got a pair, Officer Candidate Mayo-*naise!*"

Sergeant O'Hara (Lon Chaney) wooing nurse Nora Dale as Private "Skeet" Burns looks on in *Tell it to the Marines.*

Handbill for 1927's *Tell it to the Marines.*

For his role in *Tell it to the Marines,* Lon Chaney became an Honorary Marine - the first film star to do so.

Reel Marines

Steve Kanaly is a "very dangerous man" as Captain Jerome in *Wind and the Lion.*

Real-life Marine Brian Keith brilliantly portrayed President Teddy Roosevelt in *Wind and the Lion.*

Gunny Drake (Darren McGavin) giving Jan-Michael Vincent an attitude adjustment in *Tribes*.

Reel Marines

Gunnery Sergeant Thomas Drake (Darren McGavin) making sure Private Adrian (Jan-Michael Vincent) holds 'em and squeezes 'em on the rifle range. We don't want no Maggie's Drawers, all we want are fives and fours!

Adam Beach uses "the Code" to call for fire in *Windtalkers*.

Sergeant Joe Enders (Nicholas Cage) and Navajo Code Talker Ben Yahzee (Adam Beach) under fire in *Windtalkers*.

Reel Marines

Handbill for 1928 silent classic *Moran of the Marines*.

Charlton Heston as Major Matt Lewis in *55 Days at Peking*.

Lieutenant Colonel Mike Strobl (Kevin Bacon) at Chance Phelps' funeral in *Taking Chance*.

Reel Marines

Lieutenant Colonel Mike Strobl (Kevin Bacon) renders honors.

The real Mike Strobl on the set of *Taking Chance* with Kevin Bacon.

Reel Marines

Lon Chaney and Eleanor Boardman in *Tell it to the Marines.*

Patrick Swayze as Recon Marine Kevin Scott in *Uncommon Valor.*

James Franciscus in action with Tony Curtis in *The Outsider*.

Tony Curtis as Iwo Jima Flag Raiser Ira Hayes in *The Outsider*.

Reel Marines

John Garfield and Ruth Hartley in *Pride of the Marines*.

Charlton Heston, Ava Gardner and David Niven in *55 Days at Peking*.

Reel Marines

Real-life Marine PFC Gene Hackman as Colonel Rhodes.

Tex Cobb explaining to Patrick Swayze that "Using that Oriental martial bullshit on me is going to get real expensive!"

Reel Marines

Clint Eastwood as Gunny Highway (above, in *Heartbreak Ridge*), and Tom Berenger as Master Gunny Beckett (below, in *Sniper*). Both are cammied up and ready to go to work!

Reel Marines

Master Guns Beckett draws a bead on his prey in *Sniper*.

Director Luis Llosa on the set of *Sniper* with star Tom Berenger.

Reel Marines

Brian Keith as Colonel Halloran in *Death Before Dishonor*.

Colonel Halloran is wishing the Islamic terrorists torturing him would switch to something humane like water boarding.

Reel Marines

Lloyd Nolan and a very young Richard Jaeckel take cover in *Guadalcanal Diary*.

Marines prepare to hit the beach as naval gunfire preps the landing area in *Guadalcanal Diary*.

Jimmie Dundee, Eddie Bracken and William Demarest in 1944's *Hail the Conquering Hero*.

Hiep Thi Le (as Le Ly) and Tommy Lee Jones in *Heaven and Earth*.

Reel Marines

"Hanoi Jane" Fonda and Bruce Dern in *Coming Home*. The Captain played by Dern is in *major* need of a regulation haircut!

Ashley Judd and James Caviezel face the music in *High Crimes*.

Reel Marines

John Payne and Randolph Scott tangle in *The Shores of Tripoli.*

Charlton Heston and his Marines take on the Chinese "Boxers" in the 1963 epic *55 Days at Peking.*

Reel Marines

James Cagney in the 1954 version of *What Price Glory?*

Victor McLaglen, Elena Jurado and Edmund Lowe in the 1926 version of *What Price Glory?*

Marines marching to the sound of the guns in *What Price Glory?*

Reel Marines

Frank Sinatra (L) directed 1965's *None But the Brave*.

Real-life Marine Bradford Dillman didn't have to do much acting to play a Leatherneck in 1958's *In Love and War*.

Reel Marines

Ed Harris as Mercury astronaut John Glenn in *The Right Stuff*.

Robby Benson as Olympian Billy Mills in *Running Brave*.

Ed Harris as Brigadier General Francis X. Hummel in *The Rock*.

Aldo Ray, Van Heflin and James Whitmore in Leon Uris' *Battle Cry*.

Gloria Swanson tempts Raoul Walsh in 1928's *Sadie Thompson*.

Burt Lancaster makes the acquaintance of stowaway Virginia Mayo in *South Sea Woman.*

John Wayne and Claudette Colbert ride the train in *No Reservations.*

Reel Marines

Experiencing two different sides of war in 1960's *Hell To Eternity*.

Reel Marines

Burt Lancaster, Chuck Connors and Virginia Mayo go on liberty and get into trouble in 1953's *South Sea Woman.*

Reel Marines

Vamp Gloria Swanson helps a group of Marines enjoy their liberty in *Sadie Thompson*. Some things just never change!

Vincent Spano, Nastassja Kinski and John Savage in *Maria's Lovers*.

Reel Marines

Rip Torn takes charge in the 1967 film *Beach Red.*

William Bendix as Corporal "Taxi" Potts, along with Anthony Quinn as Private "Soose" Alvarez, in *Guadalcanal Diary.*

Troy (Peter Sarsgaard) and his M16/M203 enjoy a "mad moment" in the Gulf War movie *Jarhead.*

Reel Marines

Tommy Sands, Frank Sinatra, Clint Walker and Brad Dexter in *None But the Brave*.

Recruits on the grinder at MCRD San Diego in *Shores of Tripoli*.

Reel Marines

A book about Marine Corps movies would not be complete without an FMF Corpsman – in this case played by Frank Sinatra in *None But the Brave.*

Reel Marines

Robert Duvall as *The Great Santini*. "Stand by for a fighter pilot!"

Facing page: Real-life Marine veteran James Whitmore in *Battleground* (top) and a scene from *Tell it to the Marines* (below).

Reel Marines

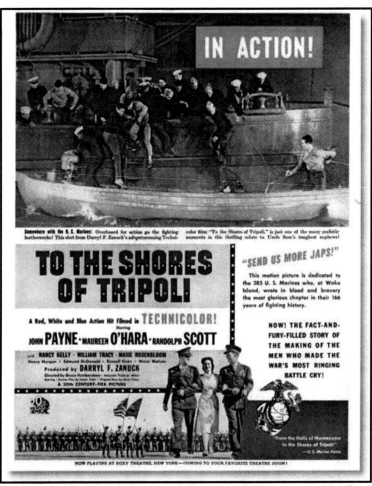

John Payne, Randolph Scott, and Maureen O'Hara starred in *To the Shores of Tripoli* - and because Pearl Harbor was attacked just prior to its release a new ending was added and the film was dedicated to the 385 U.S. Marines who were just then defending Wake Island.

CPSIA information can be obtained at www.ICGtesting.com
Printed in the USA
266511BV00006B/5/P